Georg Forster

Georg Forster

Voyager, Naturalist, Revolutionary

Jürgen Goldstein

Translated by Anne Janusch

The University of Chicago Press
Chicago and London

The University of Chicago Press, Chicago 60637
The University of Chicago Press, Ltd., London
© 2019 The University of Chicago
Published 2019
Printed in the United States of America

28 27 26 25 24 23 22 21 20 19   1 2 3 4 5

ISBN-13: 978-0-226-46735-1 (cloth)
ISBN-13: 978-0-226-47481-6 (e-book)
DOI: https://doi.org/10.7208/chicago/9780226474816.001.0001

Originally published as *Georg Forster. Zwischen Freiheit und Naturgewalt* by Jürgen
Goldstein. © Matthes & Seitz Berlin Verlagsgesellschaft mbH, Berlin 2015.

The translation of this work was funded by Geisteswissenschaften International—
Translation Funding for Work in the Humanities and Social Sciences from Germany,
a joint initiative of the Fritz Thyssen Foundation, the German Federal Foreign Office,
the collecting society VG WORT, and the Börsenverein des Deutschen Buchhandels
(German Publishers & Booksellers Association).

Library of Congress Cataloging-in-Publication Data
Names: Goldstein, Jürgen, 1962– author. | Janusch, Anne, translator.
Title: Georg Forster, voyager, naturalist, revolutionary / Jürgen Goldstein ; translated
by Anne Janusch.
Description: Chicago ; London : The University of Chicago Press, 2019. | Includes
bibliographical references and index.
Identifiers: LCCN 2018037452 | ISBN 9780226467351 (cloth : alk. paper) |
ISBN 9780226474816 (ebook)
Subjects: LCSH: Forster, Georg, 1754–1794. | Authors, German—18th century—
Biography. | Naturalists—Germany—Biography. | Ethnologists—Germany—
Biography.
Classification: LCC PT1865.F15 Z66 2019 | DDC 838/.609 [B]—dc23
LC record available at https://lccn.loc.gov/2018037452

♾ This paper meets the requirements of ANSI/NISO Z39.48-1992
(Permanence of Paper).

Contents

Prologue

Nature, a Perilous Word

When the world was still vast and undiscovered, its weight was measured in experience. Anything new was significant, from the tiniest insect in a tropical rainforest on an island that had never appeared on a map to entire continents of foreign plants and animals and peoples. Nothing was too inconsequential to merit mention in a travelogue. The world was still being discovered: maps still had blank areas and sometimes trailed off into approximations. European explorers had the sublime privilege of bestowing names on bays, coasts, islands, flora, and fauna that no European before them had set eyes on. The long-in-the-tooth world seemed young again. Foreign scents enlivened the air: nutmeg was as valuable as gold, its sought-after cloves found only on distant islands on the other side of the world. So much was unknown, unseen, unforetold. Only someone who had seen it with their own eyes—or who was skilled at invention—could chronicle the things that were new. Preserved in fusty tomes, traditional knowledge counted for little, was subject to inflation, and with each new discovery only decreased in value. One's own immediate experience was the measure of things. Like a signet ring whose design is embossed in hot wax, the impressions of a world still being discovered were imprinted on the beholder's mind. Georg Forster provided a nearly unblemished medium for such immediate experiences. He was attentive to the smallest details, while always keeping the big picture in sight, and he never hesitated to express the feelings that nature sparked in him. Forster was unprejudiced and broad-minded in meeting foreign peoples. And he articulated his experiences like no other—Georg Christoph Lichtenberg called him a "sorcerer of prose."[1] When inundated with impressions, Forster came alive. When there was nothing to see, he was weary, idle, and out of sorts. He is the subject of this book.

This book is also about a relationship, the possibility of which is so very faded that one might be inclined to dispute that it ever existed. Nature and politics may not be closely associated today, but that association is precisely what Forster was concerned with: the reality of a "natural politics" and, with it, "natural revolution" as a breakthrough in liberal self-determination by which means the "devil of feudal bondage" would be defeated. Forster spoke of "truth, freedom, nature and human rights" in the same breath.[2] He described revolution as a natural occurrence, on the order of volcanic eruptions or floods, and he spoke of it as a necessary upheaval of societal conditions. The two were not unrelated for him. Just as naturally occurring revolutions follow certain rules, the "political phenomena of one moment and one corner of the world . . . also have their cycles."[3] Forster was in search of the law that would connect the freedom found in nature with political freedom.

As a naturalist and an enthusiast, Forster's conception of nature was saturated with immediate experience. He would see, taste, smell, feel, hear, and make drawings of nature before contemplating it, and he always responded with sensitivity to it. To him, nature was neither an ideal nor something profane, but rather a staggering force that could be immediately experienced. Forster's conception of nature can be directly attributed to this immediate experience of the world, which only upon later reflection he aligned with the intellectual coordinates of his epoch.

The conditions were ripe for gaining an understanding of nature that was rooted in immediate experience. In the latter half of the eighteenth century, James Cook undertook three circumnavigations of the world.[4] Georg Forster, together with his father, Johann Reinhold Forster, took part in the second. They were at sea for three years and eighteen days. The distance traveled made up more than three times the circumference of the globe. They were the first to enter the Antarctic Circle by ship, and then they proceeded farther south than any European who had come before them. They traveled the South Seas; saw New Zealand, Tahiti, Easter Island, and Tierra del Fuego; and discovered New Caledonia and South Georgia. They came into contact with the indigenous peoples of the South Seas, unsure whether they were "noble savages" or cannibals. They observed exotic animals and brought previously unknown plants back to England. They were met with nature that was intoxicating in its beauty.

At this same time, the political upheaval that emerged toward the end of the eighteenth century permanently changed the European order. The great Revolution of 1789 proclaimed the liberty and equality of all people, breaking with the despotism that had characterized the traditional system of rule. The times seemed ripe for change. The old kingdoms were crumbling. The French Revolution was an earthquake that shook the foundations of power and caused the throne of tyranny to collapse.

Scarcely anyone was as involved in both the natural and the political concerns of the eighteenth century as Georg Forster. In Forster, the two most significant coordinates of his time converged. Indeed, "nature" and "revolution" intersect spectacularly in the thoughts and deeds of this brilliant writer, naturalist, explorer, translator, illustrator, and key revolutionary. Forster gained an inestimably rich experience of nature on his voyage around the world with James Cook. And he was at the center of political events when, inspired by the French Revolution, he declared the "Mainz Republic" in 1793—it became the first republic on German soil. No other Enlightenment thinker was on par with Forster in his experience-driven experiment of bridging the political with nature. The sparks from Forster's vision of "freedom" and the "force of nature" momentarily lit up the prospect that there could be something like a worldwide natural revolution.

Georg Forster played his part in founding a political modernity, even though his vision came to be shattered by reality. The life of this woebegone enthusiast and self-doubting optimist, this stylist and decisive actor, this highly gifted scapegrace who never quite gained a foothold in a world he knew so much about—this life ultimately unraveled, in seemingly loose strands: one was the naturalist who explored the ends of the earth with James Cook; the other was the revolutionary with the republican ethos. On the whole, these are the facets of Forster's life that tend to be discussed. Mostly, the fascination with Forster stops at the vivid descriptions of his travel experiences. Only biographers attend to both focal points, but they do so chronologically, without correctly placing the one in relation to the other.

For Forster the two belong inseparably together: perceptions of nature and of politics, the immediate experience of the world and the revolution for freedom. That is what I want to show in this book: Georg Forster was at once a naturalist and a revolutionary, suffused with the German intellectual life of his time.[5] Inspired during his voyage around the world by a glimpse of nature that could be experienced, Forster wanted to realize a new political order upon returning home—an order that struck him as natural. In pairing the terms "freedom" and "force of nature," we hold Ariadne's thread to guide us through the labyrinth of Forster's erratic life: from his early perceptions of nature to political revolution. His lifetime achievement consists not only in having circumnavigated the earth aboard a converted coal freighter, but also in having left the old world behind in order to set forth into political modernity.

A cursory glance may mistake the connection Forster draws between nature and political freedom for an exotic history of ideas. Johann Gottfried Herder, in *Ideas for a Philosophy of the History of Mankind*, tried to elevate natural history to the ranks of human cultural history. After giving an account of the earth and its position in the cosmos, as well as the animal and plant kingdoms, he turned his historical panorama to the influence of climatic conditions on human societies and illustrated the differences through a comparative look at Arctic, African, and American peoples. In doing so, he emphasized the relationship between nature and culture: "In natural philosophy we never reckon upon miracles: we observe laws, which we perceive everywhere equally effectual, undeviating, and regular. And shall man, with his powers, changes, and passions, burst these chains of nature?" Only under the Mediterranean sun could Greek culture have developed, according to Herder. "The whole history of mankind is a pure natural history of human powers, actions, and propensities, modified by time and place."[6] The supposed legitimacy of nature's influence on the development of culture does not, however, lead Herder to renounce human freedom. Climate, which he cites as the most marked influence on the evolution of culture, "promotes, but . . . does not compel, a given course of development."[7] Natural influences make people inclined to cultivate their respective cultures. But the fact remains: Humankind is free, and culture is an expression of its freedom. But humankind is not quite as dependent on nature as the Enlightenment ideal of autonomous reason would seem to suggest.

Considerations of this sort were not unusual. Jean-Jacques Rousseau already maintained, "There are . . . in every climate natural causes by which we can assign the form of government which is adapted to the nature of the climate." And Montesquieu held that the "general spirit" of a nation, the *esprit général*, was derived from the relation between laws, principles of governance, the example of the past, mores and manners, and—first and foremost—climate.[8]

Forster was of a similar mind. He believed in recognizing "from a moral point of view, the cycle of the seasons."[9] A "knowledge of nature," was, according to him, ultimately as necessary for the "cultivation of the mind and the heart" as it was for the preservation and development of the physical world.[10] In the end, "all that is moral to us has its sound basis somewhere in the physical."[11] That may sound innocuous, yet its formulation contains a radical anthropology of nature: if nature levies varying conditions on people living in different regions of the world, then how do those conditions affect the development of human faculties? Forster asks, "Would the negro who transplants his offspring to England not acquire a different set of endowments? And vice versa for the European in hot climes?"[12] Like Herder, Forster uses physical anthropology to speculate on the possibility for variation within humanity's cultivated form. The central question, however, is whether a degree of regularity can be legitimately ascertained in order to determine the influence of nature on culture. Is there a constitutional principle of the political whose discovery would help "decipher the *spirit of the present*" and also the "*signs of the future*"?[13]

In Forster's view, these were not scientific questions that went beyond the political events of his time. His anthropology of nature was conceived with pragmatic aims. To understand the conditions by which different peoples developed, Forster planned to write a handbook on natural history, of which he completed only a first draft. In it, he poses the question, "Is it not curious that among the inhabitants of the Malay islands, the feudal state was but the first step to a kind of freedom of the people?"[14] Does the climate, Forster asks, influence people's desire for freedom? Is nature a foundation for the political? Which reciprocal relationships exist between nature and politics?

It may seem absurd from today's vantage point that the idea of popular sovereignty, which would seem to challenge the condition of dependence on nature, would be groundbreaking for revolutions of the modern era. "Liberty" was the watchword of the times, and Immanual Kant, as the most formative thinker of the German Enlightenment, drew a sharp contrast between freedom and nature: There is no freedom in nature—even if in our inclinations. Freedom can be found only beyond nature. In a word, freedom is self-determination. Thus, a waking citizenry's first demand is for "civil liberty": One's private sphere of existence should be protected from encroachment by the state. Increasingly, however, "political liberty" is demanded for participation in political decisions, particularly legislative decisions. Revolution came to be understood as historical testimony of the people's inalienable right to political self-determination.

To this end, the idea of political action inspired by and codetermined by nature, of a revolution *au nom de la nature*, was quite familiar to the eighteenth century.[15] But no one would credit this epoch with a unified understanding of nature. It has been rightly noted that during the Enlightenment, "nature" was "a collective term for disparate desires for change and a battle cry against reprehensible conditions."[16] And it is worth pointing out that "proponents and opponents alike of the French Revolution could invoke nature."[17] Diderot and d'Alembert's *Encyclopedia* contained entries for "Liberté naturelle" and "Egalité naturelle," in which rising up against despotic oppression naturally appeared. Saul Ascher was convinced that "one such phenomenon of nature is political revolution." Because "the human spirit is rooted in nature," it stands in opposition to the despotic state, which acts "entirely contrary to nature's purpose, ergo revolution must follow."[18] By contrast, Edmund Burke, a decided opponent of the turbulence in France, saw everything about the revolution as having "strayed out of the high road of nature."[19] With "the license of a ferocious dissoluteness of manners," the French nation evoked "unhappy corruptions" and rebelled "with more fury, outrage, and insult than ever any people has been known. . . . This was unnatural."[20]

The revolutionaries wanted to return nature to its proper place after the decadent excesses of modern civilization. The planting of "liberty trees," decorated with bands of the French Tricolor and festooned with liberty caps, was an obvious symbol of the natural renewal of society, which was brought about by the free growth of its powers. About sixty thousand such trees are said to have been planted in the name of liberty in revolutionary France for the "reintegration of the polity into nature under the canopy of hallowed groves."[21] Proponents and opponents alike of the revolution believed nature to be on their side.

Nature—whether human nature, natural society, or nature in and of itself—proved to be flexible enough to accommodate clashing political ideas. This ambiguity over the term "nature" turned into an explosive argument. It was electrostatically charged, so to speak, with the friction of a contentious Zeitgeist. Whoever had nature on their side could lay claim to what they might not seem entitled to. Nature was turned into a legitimating power for demands and objections in all areas of life. Joseph Jubert displayed an incisive feel for the times when he noted on June 10, 1800, in *Carnet*, that "nature" had become one of the most dangerous words in the French language: "un des mots les plus dangereux dans la langue française."[22]

Even Forster rid his conception of nature of its rigorous ideas about societal development, ideas that made every political theorist nervous who linked practice and reason. In a remark that is easy to overlook, yet is of central importance, Forster wrote about the naturalness of the great Revolution in France: "Their revolution came about on its own."[23] There is a power at work that we do not have control over. For Forster, revolution is a force of nature, which inexorably breaks new ground: nature—up to and including political events—is the fate that hangs over us. In a letter dated December 29, 1793, just a few days before his death, Forster writes, "Revolution is a hurricane, who can stop it? A man who is brought to action by it does things which can only be understood in posterity, not at the moment of direness. But the slant of justice is too high here for mortals. What happens *must* happen. When the storm has passed, the survivors may relax and rejoice in the calm that follows."[24] Are passages like this merely an expression of rhetorical hubris, or are they to be taken seriously? The idea of revolution as a force of nature is not just a strong metaphor but is consistent with Forster's thinking on the divide between nature and politics as being in no way insuperable.

In Goethe this bold thesis found an advocate who could lend it credibility. He objected to the revolution, saying it was unnatural, because revolution did not satisfy his ideal of the gradual growth of fecund powers. Goethe's guiding political principles were aligned with nature's development process. And yet he also saw "natural laws" at work in the French Revolution and in the fate of the executed French king Ludwig XVI: "nature, and nothing of what we philosophers should so much like to call freedom."[25]

The conclusions Forster drew from the naturalness of political action proved to be disastrous for him. By the end, he did not understand the political world anymore. Nature had abandoned him. His mode of experience and thought had led him from an unmediated experience of nature to radicalism in political action, which in turn caused him to suffer lasting rejection and to be treated like a pariah. His own father—according to Forster's wife Therese's account—wanted "to see his son on the gallows."[26] Few remained loyal in the end. Upon news of Forster's death, Goethe wrote with sincere regret: "Poor Forster had to pay for his mistakes with his life! When he'd already evaded a violent death once before!"[27]

The debt we owe to those who reach an impasse in their thinking is rooted in the fact that we are not compelled to follow in their paths. Forster's writing is to be read biographically as documents of a developing mode of thought, which sought to reflect an unparalleled richness of experience and to transform it into political action. The chapters that follow are not concerned with his personal or private life.[28] Rather, the few decades that comprised Forster's life—he was only thirty-nine years old when he died— should be reviewed with the intention of tracing a "biography of development" that revolves around the relations between experience and action, knowledge and thought, nature and politics.

To this end, it seems advisable to me to approach Forster as Hans Stilett so brilliantly did with Michel de Montaigne, by treating "his vibrancy vibrantly, his narratives narratively."[29] It is no coincidence that Forster, a brilliant letter writer, also favored the form of a letter in his books, "because it better evokes the action for the reader."[30] Forster always had his reader in mind and sought to reach him or her by means of a prose that was as unaffected and elegant as it was lively. Because he was always attentive to its impact, his writing can seem like a dialogic continuation of his correspondence. Just as his travelogues offer "paintings of nature that do not devolve into arid nomenclature,"[31] it is lively narrative that fundamentally defines Forster's form of thinking, from *Voyage round the World* to *Darstellung der Revolution in Mainz*. To understand Forster, one must follow his trail.

Forster was an indefatigable reader, for whom intellectual continents became accessible through the years. He once noted that he had "an ocean of citations on deck" for a "thick book" he was planning to write.[32] It is no different for one writing about Forster. I would gladly step aside and give Forster the floor, sparing no quotations—he is the unread classic of German intellectual history, and it is well worth while to meet him in his own words.[33] By collaging his quotations to gather key remarks on one significant aspect—first, by drawing from Forster's extensive correspondence and journals to form the temporal context; then, by drawing more broadly from his collected works, too—the profile of his experiences and reflections should emerge.

Forster was not an actor driven by ideas. The decisiveness of his actions was commensurate with the depth of his experiences. Together, experience and action formed "the great schools of humanity."[34] His thinking sought to capture both, but one would do well not to try to encapsulate, for the sake of concision, any theory or "position" of his. Forster was neither a trained theorist nor a philosopher—he permitted himself to "philosophize in an un-philosophical way."[35] His preferred literary genre was the essay.[36] Essayistic thinking allowed him to reflect without requiring a system, to have a thrust without the rigor of noncontradiction. Ultimately, the essay provided him with a form in which his reader could be coached to maturity through a dia-logic reaction to what is presented, but without the ability or obligation to engage with unexamined assumptions.[37] Forster was a virtuoso at the eman-cipatory possibilities of the essay for sounding out ideas. Friedrich Schiller, who might not have agreed with Forster on many things, recognized it, too: "Yet, his most untenable opinions are also presented with an elegance and a liveliness that gives me extraordinary pleasure upon reading."[38]

Forster also defies any attempt to be retroactively mobilized into a frame-work. He exploited the "noble privilege" of humankind "to be inconsistent and incalculable!"[39] Yet, he was quite aware of his limits: "I am but a very small man; my nature, my education, have been so severely altered and con-strained by fate and sickliness that my abilities do not harmonize with my desires."[40] Forster understood himself to be ordinary at a time that relished its geniuses. What was extraordinary about Forster, however, was his im-mediate and multifarious experience of the world, his involvement in the course of political events, and his eloquence at capturing it all in stunning narrative language.

Essentially, he was—at the time he sailed around the world—a blank slate. He was just seventeen years old when he boarded the *Resolution*, and it took him a lifetime to process his impressions from that voyage. In retro-spect, his practical study and exploration of nature required "leisure, means, and opportunities, which only through some masterstroke should we even be able to have. My younger years were dedicated to this exhilarating busi-ness; the greatest environs for probing objective existence opened up before me: I sailed around the world. I am indebted to that voyage for the develop-ment of an endowment, which determined my course in life from childhood on, namely, endeavoring to trace my ideas back to a certain universality, bun-dling them into a unity, and thus endowing an awareness of the whole of na-ture with more life and rigorous reality within myself."[41] "Awareness of the whole of nature" did not require any specialization, any precipitous inter-est in knowledge, or any systematic grid. Consequently, what follows is primarily about tracing the "development of an endowment," which Forster, over the course of his life, carried over from his immediate experience of nature to political revolution. At the same time, this development carried him across the threshold from the old world into political modernity.

In tracing this development, I sidestep the more experience-driven intellectual biography in favor of Forster's extensive works: his seminal books, as well as his many bracing, and sometimes far-flung, essays; his reviews, his speeches, and his journals; and finally, his letters, which number more than a thousand. Together, they map his thinking and define the meridians of Forster's life.

I

Beginnings

1754–1772

:: :: :: :: ::

Johann George Adam Forster was born on November 27, 1754, in Nassenhuben, a village near Danzig, the first-born son of Johann Reinhold Forster and his wife Justina Elisabeth, née Nicolai; three brothers and four sisters followed. He was called "George" until the family moved to Germany, at which point he became "Georg." In 1765 his father took him along on a journey of several months to the Volga, on behalf of the Russian empress Catherine the Great. In 1766 father and son moved to England, where Reinhold Forster accepted a position at the Warrington Academy. His family joined him there. In 1772 Reinhold Forster was invited to accompany James Cook on his second expedition around the world. Georg went with him.

:: :: :: :: ::

Like a Blank Slate

The Greek historian Herodotus, at the beginning of the second book of his *Histories*, recounts a remarkable experiment. The Egyptian Pharaoh Psammetichus wanted to know which people were older, the Egyptians or the Phrygians. So he arranged for two newborns who were selected at random to be placed in the care of a shepherd. From that point on, the children were kept isolated; their only companions were goats, so that they could get milk. Left on their own in this manner, the children, it was thought, would develop without external influence. The children's first word would thus reveal whether the Egyptian or the Phrygian people were older. It so happened that one day the shepherd heard the two children—now two years old—calling out, with outstretched hands, *bekos*, the Phrygian word for bread. With this it became apparent to the Egyptian Pharaoh that the Phrygians' culture surpassed his own in venerability. The reliability of this experiment, as Herodotus's account suggests, stemmed from the inconvenient outcome for the Egyptians.

In this case, it is the research design that is important. Isolating the test subjects from all social ties was underscored by agrarian simplicity, because what could be learned from goats? Cultural stimuli were minimized to such an extent that the children would be able to develop from a natural state. The hope was that an immediacy would emerge from the two children if they were not encumbered by education. This model of creating an ideal situation for unhampered findings proved fascinating for modern science, because among the most peculiar findings of natural science to date was the insight that knowledge can make one blind. Modern science virtually began with a motion for censure against the traditional knowledge base. There was good reason for this position: in 1492 Columbus pushed the limits of the known world; in 1543 Copernicus displaced the earth from the center of the planetary system; in 1610 Galileo first pointed his telescope at the night sky, and the number of known stars grew immeasurably. The ancients' knowledge, passed down over generations, was being proved false. The suspicion was becoming palpable that the tomes in the libraries were full of nonsense about the world.

Ever since, all knowledge has borne the caveat that we operate with "world models" and cultivate a skeptical caution toward the latest truths and homogeneous worldviews. Above all, though, tradition has forfeited its aura of normativity. The history of science that has been handed down seems like a sequence of errors and corrections. The sheer willingness to join issue with experience as the basis for all knowledge owes its triumph to modern science. The abundance of existing knowledge can be a hindrance, however, to the acquisition of experience. It stands in the way of comprehending that which is new. What, though, if one could shelve the knowledge that has been handed down, refute it, neutralize it? No thought experiment has been more fascinating to modern thinkers than the premise of a possible tabula rasa, a new beginning without preconceptions. Since Plato's dialog *Meno*, in which Socrates demonstrates a slave's capability for mathematical learning through skillful questioning, it had been accepted that the human mind has innate ideas that need only to be roused. In the seventeenth century, this was the departure point for thinkers like John Locke, for whom the human mind resembled a blank slate, free of all ideas.[1] It is only from impressions, as David Hume formulated it in the eighteenth century, that our ideas are derived. Without experience, however, we can possess no knowledge of the world. Forster expressed this same view. He consistently defended the advantage of experience over mere ideas, which he mistrusted because, in his view, "for there to be innate ideas" was not possible. Indeed, humankind has command only over "inherited organizations and inherited susceptibility."[2] At a prominent opportunity, during the inaugural lecture of his professorship in Vilna in 1784, Forster made his position public, provoking the clergy in attendance: "Dominicans, Franciscans, Benedictines, and Jesuits were present when I demonstrated yesterday," he wrote in a letter to his publisher, Johann Karl Philipp Spener, "that mankind has no innate ideas, that its mind is material, that the whole of reason rests on received sensory impressions."[3]

It would almost seem as if nature itself engaged in an experiment when it brought Georg Forster into the world on November 27, 1754, in a remote village near Danzig. Nassenhuben, as the scattered group of farmsteads was called—Mokry Dwór in Polish—seemed isolated enough to keep the talent he was blessed with from the ballast of education. His father, Johann Reinhold Forster, was of English descent, and he attended secondary school in Berlin, studied in Halle an der Salle, and went to lectures by Christian Wolff. Despite all his ambitions, reflected by his library of twenty-five hundred scientific works, issued in wood and copperplate,[4] he became a Lutheran pastor in the provinces, without any prospect of what one might call a career. His son Georg did not attend school in Nassenhuben. He later missed the opportunity to complete his university studies, too, and thus lacked a proper education. Reinhold Forster taught Georg, but Reinhold was a dogmatic man with a tendency to quarrel.[5]

That he might sail around the world with James Cook was something Georg scarcely could have imagined as he grew up in his parents' cloistered home. The particular appeal of his observations during his three-year exploration of the world stems from the fact that he was highly gifted but had not been shaped by any educational canon that might have guided or constrained him in his observations of new and strange things. In that respect he was uneducated but tremendously capable. He himself did not consider it a flaw that he lacked the systematic outline of knowledge: "There is no wisdom from education; wisdom is merely the child of one's own experience," he later said.[6] He was a thoroughly "sensual person," so much so that his reflections, theories, and philosophical approaches routinely fell short of the force of his descriptions of nature.[7] "Nature is all the world to me," he professed.[8]

Nevertheless, he certainly saw the limits to his powers of reflection. He "neither read nor heard logic, nor metaphysics, nor natural law" and admitted to Friedrich Heinrich Jacobi, "Truthfully all that I know is not much more than mere feeling." "My complaint was always," he disclosed to Johann Gottfried Herder, "that I was yoked too soon, that I was forced to work when I still should have been learning." Such comments imply that the incredible liveliness of his descriptions stems precisely from the unspoiled quality of his immediate experience, which was virtually without prejudice, at least in intention. Forster proved himself, even in the most bewildering experiences, to be open-minded and unbiased. From Heinrich Heine comes the bon mot "Nature wanted to see how she looked, and she created Goethe."[9] We might also say, when nature wanted to be sensitively described in all her variety, she created Forster.

To be sure, this claim is unfair to the others who have explored the world and discovered nature. But the most significant German-language naturalist after Forster, Alexander von Humboldt, learned from Forster how to describe nature: not to dwell on outer appearances alone, but to portray how those appearances are reflected in the interiority of humankind. Charles Darwin, in turn, admired Humboldt, particularly the chronicle of his journey to the tropical regions of the new world. As if Forster had taken heed of the terms by which he was contracted in nature's experiment, he later wrote:

Truly, it is from darkness that man comes into the world. His soul is as naked as his body; he is born without knowledge, as he is without defenses. Bringing only a capacity for suffering into the world, he can only receive impressions of his external circumstances and allow his sense organs to be touched. The light shines long in front of him before he is illuminated by it. In the beginning he receives everything from nature and gives her nothing in return. As soon as his senses have attained greater sharpness, however, as soon as he can draw a comparison between his feelings, he goes into the wide world with his views; he sets his own terms, he maintains them, expanding them and making connections between them.[10]

Forster conceived of himself as a blank slate, willing to accept all that nature dictated to him by impression. He upheld the advantage of life and experience over theory. "Letters, formulae, and conclusions," he argued, "will never prevail over that dark and mighty drive in the young sprout to investigate through his own actions the properties of things and to ascend by experience into the wisdom of life."[11] Often it was "precisely this systematic knowledge which bars an otherwise fine mind from grasping good ideas."[12] From childhood, Forster was familiar with Carl von Linné's system of botanical and zoological taxonomy, and he expressly recognized its defining achievement of a scientific classification of nature. Increasingly, though, the system proved to be a restrictive "framework," into which Linné "fitted the things of nature." To Forster, every system that nature seeks to bring to order is only of provisional value, because "as soon as the range of vision is expanded and the viewpoint shifted," the systematic definition becomes "one-sided and half true." All systems follow from experience. For Forster, the "impartial observer" is the source of natural classification, not the definition that purports to be able to guide experience.[13] He wants "to eavesdrop" on nature, to "only record facts," then "carefully draw conclusions," thereby "banishing all exuberant hypotheses back into the narrow room" in which they are conceived by disallowing immediate experience.[14] In light of nature's exuberance, Forster is seen as being such an opponent of systems that he is distrustful of the immobilizing achievement of nomenclature: "I have adopted no particular system," he writes in *A Voyage round the World*.[15] Humankind is generally inclined "more to action than to speculation" in the world and influenced "more by feeling than by abstraction."[16]

Nevertheless, if there is a formative influence on the young Forster that can be read in *A Voyage round the World*, it is most likely that of the Scottish school, as represented by authors like David Hume, Henry Home, Adam Smith, and Adam Ferguson. Ludwig Uhlig thoroughly examined Forster's "prior understanding" of this school and singles it out as being decisive for his reflection on the circumnavigation of the world. Although it cannot be dismissed, it also should not be overstated. Uhlig himself emphasizes that Forster's reflections in *A Voyage round the World* are "always derived from observations and do not, in turn, color the chronicle."[17] Surrounded by books since childhood, Forster wanted with all his heart to remedy the limitations of his unsystematic education and always give preference to the freshness of immediate impressions over reflection.[18]

Even if the conditions of Forster's upbringing did not seem favorable, sometimes "eccentricity" is a "condition without which the highest point in the education of certain assets cannot be achieved, whereas a broadly dispersed balance of strengths is present throughout the bounds of mediocrity."[19] If Forster's remarks about the psychological conditions for the possibility of great character may be read geographically, then, Nassenhuben—this backwater in what was then the Prussian part of Poland—was eccentric enough to draw out Forster's gift for observation as a defining trait. Or, to put it another way, it was boring enough to awaken an irrepressible appetite for the world.

First Impressions from Afar

Even Reinhold Forster seemed bored to death in Nassenhuben. If one quality were found to emerge from all of the documents about the elder Forster's life, it would be his self-perception of being called to greater things. After all, he was well educated, owned scientific literature, including rare and valuable books—and had a highly talented son at his service. He actively sought out contacts, maintaining extensive correspondence with scientists and academics. And he did not shy away from change.

Catherine the Great, who ascended to the czar's throne in 1762, becoming empress of Russia, had ambitious plans. Conversant in the ideas of the Enlightenment and a correspondent of Voltaire's, she sought ways to modernize her empire. For this purpose she recruited German settlers by holding out the prospect of a golden future.[20] The reality was different: conditions were too poor and the Russian empire too vast, for the settlers to successfully cultivate the land. Fortuitously, Reinhold Forster, who had already made inquiries about the possibility of a professorship in the enlightened absolutist's empire, was given the opportunity to undertake, at Catherine's behest, a journey into the areas settled by German emigrants to get a picture of the situation and draw up an on-site report of the conditions there, knowledge of which had only been rumored. Georg, just ten years old at the time, became his travel companion. We may imagine him as being timid, since he said, "The impression of timidity and melancholy made in my youth" did not abate for some time.[21] On March 5, 1765, the two set out on the journey, which was to last one year. Reinhold Forster's pregnant wife stayed behind with the children, meagerly provided for in Nassenhuben. After a few weeks, father and son reached St. Petersburg.

What followed must have been the ten-year-old's first experience with vast expanses. In less than six months they covered a distance of twenty-five-hundred miles. From Petersburg they traveled through Moscow to the colonies on the Volga. They encountered Kalmyks, Tatars, and Cossacks and came to know stark, endless landscapes like the great Steppe east of the Volga. "In all that time, my son exercised his knowledge of nature with me," Reinhold Forster recalled.[22] Even as a child Georg was able to identify plants according to Carl von Linné's system.[23]

A wildfire they witnessed in the steppe offered a frightening spectacle. It made a profound impression on Georg; he later said of the experience: "He who knows the steppe fire in Russia will be able to imagine the terrifying speed with which fire spreads through dry grass."[24] That was when the power of nature was first manifested in Georg's life.

The German settlers' situation was abysmal. Reinhold Forster wrote a report that did not whitewash anything, but it had no effect. The journey was a financial failure, too. The ministry demurred about paying Forster the agreed-upon sum for his report. Moreover, his requested admission into the Russian Academy of Sciences was left pending. During those eight months of waiting for payment and admission to the academy, however, Georg attended school in Petersburg, where he had lessons in Latin, French, Russian, history, geography, mathematics, and statistics. This was at least a start.

Nothing lasts forever, though. Reinhold Forster left Petersburg without having received his outstanding payment and made his way with his son, not to Danzig where his wife and children were, but to England. The country of his ancestors was to bring him more luck.

The Right Place at the Right Time

The first piece of writing by Georg Forster that has been preserved is an addendum to a letter of his father's, dated November 19, 1772. No records from his childhood and youth, none of his notes or journal annotations, are available to us. We also do not know what impression London made on him.[25]

But we do know what a visit to London might have been like in the second half of the eighteenth century. Georg Christoph Lichtenberg, a brilliant aphorist and professor from Göttingen who possessed a heightened attentiveness to inner and outer worlds, described a London street in a 1775 letter from his second visit to England. "Imagine a street," he begins his chronicle of the metropolis, that has

> on both sides tall houses with plate glass windows. The lower floors consist of shops and seem to be entirely made of glass; many thousand candles light up silverware, engravings, books, clocks, glass, pewter, paintings, women's finery, modish and otherwise, gold, precious stones, steel-work, and endless coffee-rooms and lottery offices. . . . The confectioners dazzle your eyes with their candelabra and tickle the nose with their wares. . . . In these hang festoons of Spanish grapes, alternating with pineapples, and pyramids of apples and oranges. . . . All this appears like an enchantment to the unaccustomed eye; there is therefore all the more need for circumspection in viewing all discreetly; for scarcely do you stop than, crash! a porter runs you down, crying "By your leave," when you are lying on the ground. In the middle of the street roll chaises, carriages, and drays in an unending stream. Above this din and the hum and clatter of thousands of tongues and feet one hears the chimes from church towers, the bells of the postmen, the organs, fiddles, hurdy-gurdies, and tambourines of English mountebanks, and the cries of those who sell hot and cold viands in the open at the street corners. Then you will see a bonfire of shavings flaring up as high as the upper floors of the houses in a circle of merrily shouting beggar-boys, sailors, and rogues. Suddenly a man whose handkerchief has been stolen will cry: "Stop thief," and every one will begin running and pushing and shoving—many of them not with any desire of catching the thief, but of prigging for themselves, perhaps, a watch or purse. Before you know where you are, a pretty, nicely dressed miss will take you by the hand: "Come, my Lord, come along, let us drink a glass together," or "I'll go with you if you please." Then there is an accident forty paces from you; "God bless me," cries one, "Poor creature," another. Then one stops and must put one's hand into one's pocket, for all appear to sympathize with the misfortunes of the wretched creature: but all of a sudden they are laughing again, because some one has lain down by mistake in the gutter; "Look there, damn me," says a third, and then the procession moves on. Suddenly you will, perhaps, hear a shout from a hundred throats, as if a fire had broken out, a house fallen down, or a patriot were looking out of the window. . . . Where it widens out, all hasten along, no one looking as though he were going for a walk or observing anything, but all appearing to

be called to a deathbed. That is Cheapside and Fleet Street on a December evening."[26]

London was a metropolis—different from Petersburg. What's more, it served as the center of a nation that was vying for command of the world's seas, relegating France to second place and rivaling its colonial wealth. Paris may have quickly become the "capital of the 19th century,"[27] but London remained the center of world affairs.

Reinhold Forster arranged a position for his son at the trading company Lewin & Nail, where the boy had to perform servants' errands in addition to desk work in the office.[28] He had to "run around with bills and collect monies after ships had been invoiced . . . in the greatest heat, from one end of the city to the other." The work was strenuous, and he was ill suited for it. "These alternating extremes were harmful to his health," and he began showing "early signs of consumption."[29] The bustle of the big city left its mark on him. When he later described how peaceful life in Tahiti was, the conditions in England served as a stark contrast for him.

His apprenticeship may have been a brief episode, but his translation work was ongoing. Even in his youth he pursued translation—at his father's insistence, who, expecting an increase in his income, promptly made use of his son as one of his assistants.[30] Languages came easily to the boy; his style was strikingly free of dross, rich in immediate experience, and elegant. During his visit to Russia, he acquired Russian well enough that, at scarcely thirteen years old, outfitted with lexica and dictionaries, he translated a book by Michael Lomonossow from Russian into English. Although the book, *A Chronicle Abridgement of the Russian History*, was acclaimed, it did not generate money.

That one small book was not the last. The translation of *Voyage autour du monde*, by Louis-Antoine Bougainville, soon followed. Captain of a French expedition, Bougainville was the first Frenchman to sail around the world, his voyage lasting from 1766 to 1769. He described his stay in Tahiti as the high point of his travels—the South Sea island having first been discovered by the Englishman Samuel Willis in 1767. But Immanuel Kant claimed that Bougainville's glorified accounts of paradisiacal life in the South Sea represented nothing more than the "empty longing" of the European; even if one knows that what is portrayed cannot be attained, Kant writes, "it is yearnings such as these which make tales of Robinson Crusoe and voyages to the South Sea islands so attractive."[31] Forster did not agree with such

a disparaging view, complaining "that one hears philosophical minds de-
claiming against the prevailing taste in travelogue and adventure tales."[32]
Thus, before Georg Forster ever followed in the footsteps of Bougainville
and traveled around the world himself, he first read about this foreign paradise on the other side of the world, translating the foreign impressions from
a circumnavigation. It is surprising—and speaks to the power of his perceptions—that he did not lapse into Bougainville's clichés. And the travelogue
did make an impression on Forster. Why wouldn't it? A voyage around the
world was the epitome of breaching all known boundaries.

The six years that the two Forsters spent in England were confining.
Reinhold Forster started a new position at the Warrington Academy in
northwestern England. It was a modest position, but it enabled his wife and
children to join him from Nassenhuben. Georg was released from his position at Lewis & Nail in London, and his father enrolled him as a student at
Warrington. Georg was the 159th student since the founding of the academy
in 1756. He studied a bit of math, physics, morality (and "the truth of the
Christian faith"), and French. Above all, though, he took to drawing.[33]

His father, who soon became dissatisfied with his position and estranged
from the academy's rector, would not think of letting his talent languish in
the provinces indefinitely. He successfully petitioned for admission to the
Royal Society, the leading academy of scientists.[34] This placed him in the
right place at the right time: James Cook was planning his second circumnavigation of the globe. Joseph Banks had been engaged as chronicler of the
voyage, but he required extensive accommodations on the ship to enhance
his comfort during the long voyage; otherwise his seaworthiness would be
adversely affected. Cook and Banks reached a stalemate and parted ways.
Time was of the essence. When Reinhold Forster was approached, he accepted without hesitation and was able to negotiate taking his son along on
the voyage. Georg was not asked. "I was finally forced to go into the world,"
he later said, "under circumstances that resulted from a situation that I did
not choose for myself. I could not and was not to breach this arrangement
and so I bowed my head to fate."[35] He was just seventeen, barely old enough
for a voyage of this kind, but just young enough for a genuine view of the
world.

II

Views of Nature

The Voyage around the World, 1772–1775

:: :: :: :: ::

Reinhold and Georg Forster accompanied James Cook on his second circumnavigation of the globe, from July 13, 1772, to July 30, 1775—three years and eighteen days. Cook was commander of the HMS *Resolution*, with a crew of 117 men, including the two Forsters, while the HMS *Adventure*, with a crew of 83 men, was under the command of Captain Tobias Furneaux. The purpose of the expedition was to make a reconnaissance of Antarctica to determine whether the southern hemisphere contained another great continent (Terra Australis Incognita).

They set sail from Plymouth, England. After stopping in Maderia and the Cape Verde Islands, they crossed the equator on September 9, 1772, and spent a few weeks in Table Bay, South Africa. On January 17, 1773, they crossed the Antarctic Circle. Over the course of several weeks, Cook attempted to push as far south as possible. In February 1773 the two ships lost sight of each other, and after meeting up again at a preordained point along the coast of New Zealand, they became separated for good on October 29, 1773. The *Adventure* returned to England on July 15, 1774.

Facing the Antarctic winter, Cook decided at the end of February 1773 to turn north. But several more weeks in the Antarctic Ocean passed before he gave the order to sail northeast. They wintered in New Zealand, reached

Tahiti on August 16th, and spent the next few months exploring the islands of the South Pacific—Huahine, Raiatea, and the Hervey (today Cook) and Tonga Islands. In November 1773 they returned to New Zealand, to embark from there on their second Antarctic voyage. On January 30, 1774, the *Resolution* reached 71° S latitude (71° 10′). No European had come this close to the South Pole before. An immense ice field kept them from continuing.

In March 1774, they reached Easter Island. In April they returned to Tahiti, by way of the Marquesas Islands and the Shetland Islands. From Tahiti they headed for Fiji, sailing a course through Huahine, Raiatea, and the northern Tonga Islands. After passing through the New Hebrides and countless additional islands—among them Mallicolo (Malekula) and Tanna—they discovered New Caledonia. From New Zealand, they crossed the Pacific without making a single stop, rounding Cape Horn at the end of December, and as they continued on course for South Africa, they discovered South Georgia and the Sandwich Islands. Upon their arrival back in Table Bay, South Africa, on March 22, 1775, they found that their timekeeping, which indicated March 21, was off by one day. On June 11, 1775, they crossed the equator once again and arrived in England on July 30.

They had covered—taking all courses into account—a distance of more than three times the circumference of the earth. During the voyage, four crewmembers died on board the *Resolution*. In New Zealand alone, ten seamen from the *Adventure* were killed.

Forster's *A Voyage around the World* appeared in 1777, and the translation, *Reise um die Welt*, was published in two volumes in 1778 and 1780 by Haude & Spener in Berlin.

:: :: :: :: ::

Patterns of Perception

Forster's lack of preconceptions at the start of the circumnavigation and his independence from any educational canon gave him an unaffected view of nature. Patterns of perception can be read in his descriptions of these views, which Forster developed during the three-year voyage. He processed impressions that exceeded all expectations. He vividly described the months-long experience on rough seas, the storms that tossed the ship, the reefs and icebergs that threatened to shipwreck the expedition. Sailing the vast oceans was the quintessence of distance, and it made the Eurocentric point of view of the explorer fade in a very physical way. The animal and plant world was in every respect staggering and exotic. The voyage alternated between extremes, the extended exploration of Antarctica amid ice and fog followed by the paradise-like Pacific. And the contact of Forster and the crew with foreign peoples ranged from the cannibals of New Zealand to the unfortunate "creatures" of Tierra del Fuego and Easter Island, to the inhabitants of Tahiti, whose grace and poise were, to Forster, reminiscent of the ancients.[1]

Forster's way of thinking was directly shaped by these impressions. In an anthropological respect, his thinking strives to mediate between unity and difference. Thus he is able to emphasize the unity of the human race, which asserts itself in spite of concrete differences in the immediate experience of peoples around the globe. Forster seeks proof "that men in a similar state of civilization, resemble each other more than we are aware of, even in the most opposite extremes of the world." For him, "the passions of mankind are similar everywhere; the same instincts are active in the slave and the prince; consequently, the history of their effects must ever be the same in every country."[2] Yet Forster discovers a "relative morality," a cultural relativity that yields different moral systems of norms, whose differences are worth observing.[3]

At the same time, his immediate experience of nature, which proved to be so staggering, caused his political thinking to become saturated with experience to the point of metaphor. He later compared revolutionary political processes to volcanic eruptions. What does it mean, then, for Forster to come face to face with an active volcano? That the backdrop for his thinking was shaped "realistically" rather than "idealistically" may have precipitated his willingness to become radicalized, rather than become merely a sympathizer of political revolution when his interests turned to politics a few decades later. Where others' ideas flagged or were kept in check by deploying caution and reason, Forster devoted himself to the illusion that what he had experienced on the other side of the world could now become reality in Europe.

Forster experienced nature as a mighty kingdom wrestling with its own forces. This is why, with clear political overtones, he dismisses the Aristotelian teaching to strive for the mean. "Many philosophers have sought zealously to find a restful mean between the two extremes, and many have thought they had found such a complete equilibrium—but this is also the absolute quiet of death." In the political realm, it is far more acceptable to yield to those forces that are also at work in nature. In this way, "violent movements from one extreme to the other can prevent a dangerous stagnation in the great course of humanity."[4]

The forces at work in both cases possess a destructive-productive force.

Forster writes in *Cook, the Discoverer*, "The course of so many revolutions, all of which resemble one another, no matter how the local conditions or the time may change, seemingly destroys the system of idealism by showing it to be founded on a baseless hypothesis."[5] The power and impact of revolutionary processes resemble the violence of ocean waves, which, on the beach at Dünkirchen in April 1790, Forster finds frightening as they "crash with unrelenting severity, washing away everything before them that contains life in the tides."[6] The leaders of revolution, too, "are in the whirlpool and do not steer it," and although every individual tries to gain "advantage from each wave," all are subject to the "violent momentum, which carries everything away."[7] Is this consonance between his perception of nature and the nature of politics merely metaphorical?

Hardly. It should be pointed out, with due caution, that Forster's patterns of perception affected his approach to political phenomena during the French Revolution. This mode of thinking forms the virtual key to how Forster structured *Ansichten vom Niederrhein* (referred to hereafter as *Views of the Lower Rhine*). By employing a proto-ethnographic perspective to cope with the profusion of impressions, he applied his intuitive method of writing, developed during the voyage around the world, to the complex political situation in Europe after the French Revolution.

First, Forster foregrounded nature: the Antarctic ice and the vast sea, the tropical landscapes with their exotic animals and plants, far-flung islands and foreign shores, the world of the Pacific and the Atlantic oceans. But the many facets of human nature came to interest him, too, how they are reflected in physiognomy: faces that resemble those of apes, the beauty and repellant ugliness of the human creature among the Tahitians and the inhabitants of Terra del Fuego. He marveled at the variety of skin tones, to which the fairness of the Europeans became the exception. Their societies, too, Forster eventually came to understand as natural; he got to know them only poorly but grasped them quickly during the brief stays here and there. A people's character and their culture could be attributed to the influence of nature.

It makes little sense to me to follow the chronology of the travelogue. At nine hundred pages, *A Voyage round the World* is an inexhaustible source that can hardly be condensed or summed up. One should resist the attempt to present a handy abridged version, something that Forster himself had no intention of writing or was in no position to write. The scope of the book and the stamina it demands of the reader reflect exactly the temporal and spatial scale of this epic expedition. Thus, it seems more promising to use a series of individual aspects to draw out those contrasts and features that manifest Forster's perceptual patterns, which in turn structure *A Voyage round the World*. Discussion will turn to the hardships on board the ship and the joys of a simple life in the South Seas; to the ocean, the ice, and the tropics; to the virtually untouched peoples of the Pacific and their seamier side; and ultimately to the impressions that enabled a new perspective on old Europe. This richness of worldviews, which can hardly be contained between two book covers, cannot be uniformly organized. The abundance of what he presented refuses any attempt to be reduced to a quintessence. Forster was well aware of this, and it is why he did just as he was advised: he told the story of his voyage around the world.

A Well-Told Tale

Georg Forster is a storyteller. His prose is committed to immediate experience, and it always has the reader in mind. His *Voyage round the World* is not dry reportage, nor is it a chronicle of nautical facts. He conceives of his travelogue as literature, on par with the novel but without the license to invent. No flights of fancy were required—his immediate experience of the world was fascinating enough. He wanted to recreate for the reader what he had experienced those three years through unaffected language, evocative and modern, precise but not academic: "The degree of pleasure which may result from the perusal of a work, depends not only upon the variety of the subject, but likewise upon the purity and graces of a well-told tale to a lame and tedious narration." Forster writes about his voyage around the world for a readership that wishes to be not only informed but entertained. He believes it is "the events of the voyage, the dangers to the voyagers, the hardships suffered and the behaviour of the inhabitants of far-distant countries, in a word, the action which engages the passion of the reader and makes a book of travels supremely interesting."[8]

In the eighteenth century, the connection between literature and science, epic prose and research, was well intact, having been fostered during the early modern period. As early as 1609, Johannes Kepler wrote a fantasy narrative about a trip to the moon; Galileo defended his Copernican worldview in *Dialogue concerning the Two Chief World Systems*; and Bernard de Fontenelle followed suit with his entertaining yet erudite *Conversations on the Plurality of Worlds*. Even Goethe was still planning in 1781 to write a "novel about the universe." So it seemed only natural, then, to write a "well-told tale" about that spectacular voyage whose eyewitness, gifted as he was with language, was Forster. Later, Forster even entertained the idea "of playing into the hands of women by teaching natural history to children in an inviting and comprehensible way."[9] He wanted to reach a broader readership than just an audience of experts.

Forster wanted to awaken within the reader the very feeling that had been sparked in him by contemplating nature. "For me, feeling is greater than knowing,"[10] he confesses, and "living, fruitful abundance [is] quite distant from cold pedantry and childish trifles. It doesn't merely enrich the memory, it doesn't merely awaken the powers of thought, it doesn't merely spark the visual drive of the imagination—it also speaks to the heart, impregnating it with the complete spirit of beauty and goodness."[11] He avoids the undergrowth of facts to span the whole of nature with a narrative arc. "I have sometimes obeyed the powerful dictates of my heart, and given voice to my feelings," he admits. Stylistically, however, "without attempting to be curiously elegant," he "aimed at perspicuity,"[12] and his travelogue is proof that the two are not mutually exclusive.

To that end, Forster clearly wanted to break away from "the common travelogues."[13] The "philosophical history of the voyage" that he provided was "free of prejudice and vulgar error." The "discoveries in the history of mankind, if not natural history" were portrayed "without any adherence to fallacious systems, and upon the principles of general philanthropy." In short, *A Voyage round the World* attempted to offer an account of the sort "which the learned world had not hitherto seen executed."[14]

Particular attention must be paid to the portrayal of people. Forster's *Voyage round the World* is different from Cook's travelogue in its anthropological basis and its interest in ethnography. "Doubtless no topic is of greater interest to mankind than man himself in all his manifold shapes, stages of development, and conditions of time and place," Forster writes in *Cook, the Discoverer*.[15] Columbus, in characterizing the natives of the land he found, which he mistook for India, as "naked," employed a metaphor for "a lack of culture and religion." Forster, by contrast, is interested in recording foreign cultures. "I have always endeavored," he writes, "to throw more light upon the nature of the human mind" and, in so doing, to speak with praise or censure, "with neither attachment nor aversion to particular nations." It is from this intention that the spirit of Enlightenment speaks, "accustomed to look on all the various tribes of men, as entitled to an equal share of my good will." But Forster is not naive; he would like to reveal to the reader "the colour of the glass, through which I looked," since he would also like to understand it himself.[16] He accounts for his own position, from which he regards these exotic cultures, and consequently comes to know his own perspective. "I was born, raised, and through entirely no fault of my own, a crease was pressed into my way of thinking, a direction given, entirely unremarkable; and lo! I now think this instead of that."[17] Even the most unbiased ways of looking at the world are biographically and culturally colored, every judgment imbued with experience. Hence, "two travelers seldom saw the same object in the same manner, and each reported the fact differently, according to his sensations, and his peculiar mode of thinking." At least he was aware that the colored glass through which he looked had "never clouded" his sight.[18]

Thus, Forster's goal is to deliver a "narrative with the most scrupulous attention to historical truth"—a true story rather than any theory, an account of experience rather than an accumulation purely of facts: "To be a mere compiler . . . that I cannot do." Despite his esteem for the observation of individual things, Forster seeks to avoid grappling with "the million-fold hydra of the empirical," as Goethe called it.[19] It is without anecdotal embellishment that Forster writes of impending shipwrecks, storms, ceaseless voyages on the open sea, and treacherously close icebergs. What might be mistaken for narrative ornament—entertaining, but of no scientific importance—would be more accurately viewed as an exact documentation of the conditions for his experiment: to behold the world is to expose it. Each glimpse to the ends of the earth required having put one's life at risk. On the first circumnavigation of the earth, which Cook undertook from 1768 to 1771, twenty-six men died of illnesses and two seamen had fatal accidents. All of their stories are woven together, as with Ariadne's thread, through the "labyrinth of human knowledge."[20]

Even in his later remarks, Forster defends his avoidance of scientific-academic hermeticism. "What is this wretched consequence of methodical or pedantic attitude," he exclaims in a letter to Christian Gottlob Heyne in January 1792, "to which our professors so easily fall prey. They want only to glean, to collect. Their wider perspective gets muddled, and their moral eye becomes accustomed to a murky myopia." "Nothing about that feeds the spirit. Withered letters can provide no nourishment, if receptiveness is not delicately preserved through the pride of conscious awareness."[21] The addressee, Forster's father-in-law, held the chair in classical antiquity at the University of Göttingen, and, as the most senior librarian, he was familiar with the writings of Vergil and Pindar—Forster clearly had him in mind when defending his feelings. For Forster, "reason without feeling" was "calamitous." He did not want to submit himself to the "*tyranny* of reason,"[22] the "systematic tyranny of the soul,"[23] and be forced into a "narrow framework" of thought. He is not opposed to the enlightened power of reason but to its one-sided absolutism, as scourged by Schiller in his *On the Aesthetic Education of Man.* Forster's high regard for observation, experience, and empathetic empiricism thus serves as a necessary corrective to the Enlightenment's impending one-sidedness, which "elevated cold reason into an all-beloved idol at the expense of feeling." Narrative is the linguistic medium in which these experiences are stored, without having to be forced into a system. Although it is true that no description can replace the "living impression which we gain through our own senses," part of the responsibility of language is to deploy its narrative potential in the service of constructing an account of what was experienced, to satisfy the "value of the fleeting moment."[24]

Accordingly, Forster writes quickly and can hardly be called pedantic. "If we didn't have anything else to do," he writes to his publisher Spener, "then, we would spend so long polishing our writings that a rule of grammar could be abstracted from each line, and entire academies should marvel at us; this alone would cause the ultimate aim of education to fall by the wayside." He did not have time for that, not in his days or in his lifetime. What Forster seeks to capture in his writing is that fleeting moment of sensory fulfillment through immediate experience. His descriptions of nature are an echo of completed experiences, written with full awareness of the inevitable: "Bliss! And tomorrow we may die!"[25]

Writing was thus sometimes a race against time. One does not see in *A Voyage round the World* the haste with which it was written. He "worked [him]self sick" on it, and over the course of its writing, he became "extremely miserable from bilious maladies"—*bile* being the French word for gallbladder—"tremendous headache, chills, and diarrhea, very nearly a skeleton."[26] The elegance of the language redeems the memory of the effort required to put the voluminous book on paper and then—admittedly not without help—to translate it into German.[27] "With which spirit, with which mood does the travelogue progress, when everything around us stands sorrowful, bleak, and barren, you can guess," he comments to his publisher. He signs the letter, "Until I come to a miserable end."[28]

Time and again he became painfully aware of the inadequacies in his education. While writing his essays, books, and lectures, while translating—a sometimes "highly intolerable work"[29]—he became aware of all that he "had and *had not* learned," and this was why he needed "threefold and tenfold the time than it should take others" to complete his work. He was always running up against unfavorable circumstances in his life. "Time fails me everywhere," he groans to Jacobi; it goes "to wretched, worthless matters."[30]

Even though his stylistically brilliant *Voyage round the World* was a high point in travel literature, Forster remained a self-doubting author, convinced he had "written nothing right yet." The insecurity was deep-seated: "I alone have an illness, a doubt in my own abilities, which only seldom lets me believe that I am capable of lifting myself above mediocrity." The sovereignty that should have come with success failed to materialize. "I still have much to learn before I can write a good book; and if I don't soon begin, I will not even be in any position to study, and my best years will be lost," he lamented some years later. He would need to be able to read undisturbed for three years, "in order to eventually become a writer through diligent study, including study of the Ancients."[31]

On the "turbulent seas of writing" Forster roved about, not least because he needed money. "Shackled to the quill, just as in the galley," was how he had put it since his youth when appraising his own contributions as an author and translator. "Writing is not even my inclination, rather it is an obligation towards my creditors," he complained bitterly in another instance.[32] This was said with respect to his publisher, and hence perhaps not without ulterior implication. Still, this avowal is important to our understanding of Forster as an author. He did not see himself as an "original genius," which, since Robert Wood's 1769 *Essay on the Original Genius and Writings of Homer*, had served as the exemplar for the epoch. In Forster's view, it was "not every man's concern to say, or write, or invent something new." Geniuses are rare. Forster knew to appraise his abilities self-critically and without arrogance. When he experienced something, he wrote about it. Should the experiences fail to materialize, his quill dried up. He was not necessarily inventive, nor he did possess the fertile imagination with which to create literary worlds. "I don't know if I'll ever write anything of my own again," he once said. All of his writing—at least, anything of importance— was dependent upon his immediate experience. When he realized that the process of writing *A Voyage round the World* had caused the vital impetus for his writing to cease, the idea of repetition became promising: "I can well see how I might be of greater use to the world if I were to embark on *another* great voyage, view it objectively, and could chronicle what I saw in good faith."[33] Although it was not to be, the desire points to the source of Forster's art of storytelling: experiencing the world.

The Sea

Odysseus's journey is understood to be punishment for a sacrilegious offense. For years Odysseus was condemned to being helplessly subject to the might of the sea, unable to return to the home he longed for because he had overstepped the rules of ancient cosmology: his odyssey was the consequence of a border violation. To venture out to sea is to leave the human realm of life behind. "Among the elementary realities we confront as human beings," writes Hans Blumenberg,

> the one with which we are least at ease is the sea—with the possible exception of the air, conquered later on. The powers and gods responsible for it stubbornly withdraw from the sphere of determinable forces. Out of the ocean that lies all around the edge of the habitable world come mythical monsters, which are at the farthest remove from the familiar visage of nature and seem to have no knowledge of the world as cosmos. Another feature of this kind of uncanniness is that myth assigns earthquakes—since time immemorial incontestably the most frightening of natural occurrences—to the sea god Poseidon's realm.[34]

Going to sea necessarily means leaving the solid ground beneath one's feet. Metaphorically speaking, the sea epitomizes imponderability.

The history of successfully expanding human territory includes the feat of integrating the Mediterranean into humanity's cultural domain during antiquity.[35] The Romans named the Mediterranean *mare nostrum* ("our sea") not only to denote their claim to power over its territory, but to express intimacy with the sea. Mediterranean trade, the fishing industry, even the naval wars, were attempts at "humanizing" the sea, which quickly forfeited power to the other side of the Mediterranean catchment area. Sailing the ocean was long deemed formidable. The "Pillars of Hercules" (the Strait of Gibraltar) represented the ends of that world, which people were not advised to roam past. Beyond that point awaited the unknown, and anyone who was not afraid of it was guilty of recklessness. Even the Old Testament story of Leviathan, the sea monster that God alone could command, references the threshold that is crossed when leaving behind the altogether fathomable Mediterranean.

The young Georg Forster did not have to know any of this to feel the singularity of the moment when he departed Plymouth Sound on July 13, 1772. Not least, the seventeen-year-old was leaving behind his mother and his siblings, without any certain prospect of seeing them again: "I took a parting look on the fertile hills of England, and gave way to the natural emotions of affection which that prospect awakened"—it brought him to tears. The lighthouse in Eddistone, built on a boulder in the middle of the sea, served as the last witness to the old world he may have beheld until it, too, disappeared beyond the horizon. It was a voyage into uncertainty. Not only was his own return anything but certain, but reuniting with those left behind seemed uncertain, too. "All those who had left behind them relations and parents," Forster writes toward the end of the voyage, "were apprehensive that they had lost some of the number during their absence; and it was more than probable that this interval of time would have dissolved many valuable connections, diminished the number of our friends, and robbed us of the comforts which we used to find in their society." Once they reached open sea, though, he brushed aside these fears. "The beauty of the morning, and the novelty of gliding through the smooth water" won the upper hand and "dispersed the gloominess of former ideas."[36] What other feelings can be confessed upon departing into the unknown when there is a world to discover and one's own sensitivities to be set aside?

The ocean did not stay calm for long. There were heavy swells, and Forster, who had previously sailed only the Baltic and North Seas, felt wretchedly seasick. Even on later voyages he was beset by nausea: "The Seasickness has something dreadful in it; it made me indifferent to everything in the world."[37] Ultimately he was able to get used to the swells. But the peril of Cook's expedition was at no point lost on him. They had already narrowly avoided catastrophe in Plymouth Sound. The fully loaded *Resolution* had become unmoored and started drifting, in danger of being dashed against the rocks. At the last moment the seamen succeeded in hoisting the sails and heading off catastrophe. "We shall, in the course of this history, find frequent instances of impending destruction, where all human help would have been ineffectual, if our better fortune had not prevailed under the superior direction of HIM, without whose knowledge not a single hair falls from our heads." In retrospect, the circumnavigation may very well have been under the "guidance of divine providence." For the time being, though, Forster was shown in a rather unsettling way how an undertaking of this sort was reliant on "a higher power" and did not rest in human hands alone.[38]

Forster's descriptions of the storms their ship endured during its three-year voyage read like the depiction of a struggle, as though the sea were punishing intruders for a border violation, playing cat and mouse with them. "The ocean about us had a furious aspect, and seemed incensed at the presumption of a few intruding mortals," Forster writes in January 1774. During this storm, "at nine o'clock a huge mountainous wave struck the ship on the beam, and filled the decks with a deluge of water. It poured through the skylight over our heads, and extinguished the candle, leaving us for a moment in doubt, whether we were not entirely overwhelmed and sinking into the abyss."[39] As a plaything for the waves, the *Resolution* often just barely escaped shipwreck.

In October 1773, Forster depicts a storm off the coast of New Zealand as though the sea had decided to finally dispose of the explorers. In a longer passage, he paints the drama on the sea by presenting it to the reader in breathless, shimmering sentences:

> Though we were situated under the lee of a high and mountainous coast, yet the waves rose to a vast height, ran prodigiously long, and were dispersed into vapour as they broke by the violence of the storm. The whole surface of the sea was by this means rendered hazy, and as the sun shone out in a cloudless sky, the white foam was perfectly dazzling. The fury of the wind still encreased so as to tear to pieces the only sail which we had hitherto dared to shew, and we rolled about at the mercy of the waves, frequently shipping great quantities of water, which fell with prodigious force on the decks, and broke all that stood in the way. The continual strain slackened all the rigging and ropes in the ship, and loosened every thing, in so much that it gradually gave way and presented to our eyes a general scene of confusion. In one the of the deepest rolls the arm-chest on the quarter-deck was torn out of its place and overset, leaning against the rails to leeward. A young gentleman, Mr. Hood, who happened to be just then to leeward of it, providentially escaped by bending down when he saw the chest falling, so as to remain unhurt in the angle which it formed with the rail. The confusion of the elements did not scare every bird away from us: from time to time a black shear-water hovered over the ruffled surface of

the sea, and artfully withstood the force of the tempest, by keeping under the lee of the high tops of the waves. The aspect of the ocean was at once magnificent and terrific: now on the summit of a broad and heavy billow, we overlooked an unmeasurable expanse of sea, furrowed into numberless deep channels; now on a sudden the wave broke under us, and we plunged into a deep and dreary valley, whilst a fresh mountain rose to windward with a foaming crest, and threatened to overwhelm us. The night coming on was not without new horrors, especially for those who had not been bred up to a seafaring life.[40]

The danger only increased when they lost sight of the *Adventure*. For safety reasons, Cook had started out with two ships, but the crew of the *Resolution* was forced to realize "that from now on, we were obliged to proceed alone on this vast and unexplored expanse." Their hope of a triumphant homecoming made them forget how improbable a return even was. It was no trivial matter for them to embark on a trip around the world in a ship that had been converted from a coal freighter. Any damage to the vessel only increased their risk of becoming unable to maneuver. After departing New Zealand, they discovered they had sprung a leak, "but it gave us very little uneasiness, as the water in the pump-well encreased only five inches in eight hours," Forster writes, understating the risk in retrospect. A storm was once so strong that Captain Cook fastened a copper chain to the masthead, draping it over the ship's rail into the sea. When "a terrible flash of lightning appeared exactly over the ship, . . . the flame was seen to run down along the whole length of the chain. A tremendous thunder-clap instantaneously followed, which shook the whole ship, to the no small surprise of both the Europeans and Tahetians on board."[41]

The swells were sometimes so high that the ship would tilt forty degrees.[42] Once they were pushed toward an island whose coast "was of a great height, rocky, black, and almost perpendicular." Because there was no wind, they were subjected to the surge of the waves: "The current was so strong, that hoisting out our boats would scarcely have availed us any thing. The ship's head, her stern, or her broad-side, were by turns directed towards the shore, on which we heard the surf breaking with a much more dreadful sound than it had ever had before, when unconnected with the ideas of immediate danger; at last we fortunately drifted clear of the point at short distance."[43]

What in the twentieth century would be called an "existential boundary situation" was a recurring experience on the three-year voyage around the world. Apart from the health of the crew, the Achilles heel of the expedition was the ship. Their return depended on the vessel's remaining intact. It caused a panic when one evening the alarm was given:

> Towards ten o'clock, we were most dreadfully alarmed by a fire in the ship. Confusion and horror appeared in all our faces, at the bare mention of it; and it was some time before proper measures were taken to stop its progress: for in those moments of danger, few are able to collect their faculties, and to act with cool deliberation. The mind which unexpected and imminent danger cannot ruffle for a time, is one of the scarcest phaenomena in human nature; no wonder then, that it was not to be met with among the small number of persons to whom the ship was entrusted. To be on board of a ship on fire, is perhaps one of the most trying situations that can be imagined; a storm itself, on a dangerous coast, is less dreadful, as it does not so entirely preclude all hopes of escaping with life. Providentially, the fire of this day was very trifling, and extinguished in a few moments. Our fears suggested that it was in the sailroom; but we soon found, that a piece of Taheitee cloth, carelessly laid near the lamp in the steward's room had taken fire, and raised a quantity of smoke, which gave the alarm.[44]

The elemental force of violence seemed unfettered when, in May 1773, they observed four waterspouts, the closest of which was just three miles from the ship. Just a few hundred yards away, the sea was churning "in violent agitation," as Forster reported: "The water, in a space of fifty or sixty fathoms, moved towards the centre, and there rising into vapour, by the force of whirling motion, ascended in a spiral form towards the clouds. Some hailstones fell on board about this time, and the clouds looked exceedingly black and louring above us." They could observe how "the water hurled upwards with the greatest violence in a spiral." When the last waterspout dissipated, they saw lightning, but heard no thunder. "Our situation during all this time was very dangerous and alarming," which "made our oldest mariners uneasy."[45]

Experiences of this sort left their mark on Forster. It is not an exaggeration to say the sea brought him to the limits of his mental and emotional stamina. The sea became for him the embodiment of death and destruction. He survived.

Of all the dangers posed, at least they were unharmed by sea creatures.

Among their most impressive experiences with the foreign animal world was the sighting of some thirty whales, one of which came within two hundred feet of the ship. In spite of their size, Forster marveled at how they "sometimes fairly leaped into the air, and dropped down again with a heavy fall, which made the water foam all around them."[46]

There was a later echo of this boundary situation. In April 1790, Forster took a trip with Alexander von Humboldt, during which he saw the ocean in Dünkirchen again for the first time in twelve years. "The sight awakens myriad thoughts!," he wrote to his wife. The contemplation of the open sea shook him, triggering premonitions of death. He was worldly-wise, yet afraid in the face of this boundless element, as if the experience of having sailed to the edges of life as a young man might now overpower him. "I will not be able to describe to you what happened within me," he writes in chapter 20 of *Views of the Lower Rhine*, in an effort to relate his feelings to his wife and the reader. "Abandoning the impression that this sight made on me, I instinctively sank back into myself, as it were, and the image of those three years, which I spent at sea, and which determined my entire destiny, loomed before my soul. The vastness of the sea seizes the viewer, darker and deeper than that of the starry sky. There, on that silent, unmoving stage twinkled endlessly insoluble lights. But here, nothing is essentially distinct; a vast body, and the waves merely fleeting phenomena." The waves "arise and tower up, they froth and disappear; they are engulfed again by the vastness. Nowhere is nature more terrible than here in the unrelenting severity of its laws; nowhere does one feel more clearly that, set against the whole of matter, the wave is the only thing that passes through a point of separate existence, from nonbeing back into nonbeing; yet, the whole rolls forward into immutable unity."[47] For one brief moment, the individual is a plaything for nature. While standing along the edge of the sea, Forster is conscious of death and perhaps overcome by the insight that just as nature gives all, it takes all, too. It is sacred and cruel. Against the forces of nature, we cannot compete. He tells his wife he has no intention of "going back to sea, not until I have nothing to lose."[48]

Distances

Even in the sixteenth and seventeenth centuries, it took one to two weeks on average to cross the Mediterranean from north to south. For passage from west to east, two to three months.[49] Cruising speed was not something that could be calculated, as storm and calm alike could lay waste to any plans. Because the open sea was to be avoided, seafaring was conducted along coastlines, *costeggiare*, the practice was called: *navigare a poca distanza dalla costa.*[50]

The advances made after Columbus's journey of discovery, namely ocean crossings, indelibly changed seafaring. Every voyage of this kind opened up the world. Cook's second circumnavigation, as Forster writes, was "an unexampled navigation." Unlike Columbus, they sailed against the sun: "Among all journeys around the world ours is indeed the first which was directed from west to east."[51] Upon their return to Table Bay in March 1775, they determined that they had gained a day.[52]

Sailing the world's oceans was not just physically strenuous. The psychological strain sometimes weighed more heavily on the seamen and their fellow travelers than any physical deprivation. "Our navigation, which for nine weeks past has been out of sight of any land," Forster noted during the first year of the voyage, "began to appear dull and tedious, and seemed to be distressing to many who were not used to an uniform recluse life on board a ship, without any refreshments or variety of scenes." The weeks spent on open water were defined by emptiness. During their journey of more than three years, "the sum total of all the days which we had spent on shore . . . did not amount to more than one hundred and eighty, or about six months." Monotony was the hallmark of the "solitary hours of an uniform navigation."[53] Forster worked out that at one time they had been at sea for 103 days straight without seeing land; another time, four months and two days, and they spent a total of two years and nine months in the southern hemisphere.[54] In the endless expanses of the Atlantic and Pacific oceans, and in their attempts to press forward into Antarctica, they could feel themselves wasting away. "Repeated calms rendered our course very tedious." The ocean, to go by their experience, proved to be just about interminable. "It is obvious that to search a sea of such extent as the South Sea, in order to be certain of the existence, or non-existence of a small island, would require many voyages in numberless different tracks, and cannot be effected in a single expedition."[55]

Tahiti was on the other side of the world, and since Forster was doubtful "whether another ship would be sent to Taheitee again," every glimpse of this foreign world was singular. Their immediate experiences were exclusive—and Forster knows why. If "ever a voyage should be undertaken again" to sail around the world, he could hardly hope to find himself on board again. In December 1773, when they reached the antipodes of London—the point diametrically opposite the English metropolis—they were "the first Europeans, and I believe I may add, the first human beings, who have reached this point, where it is probable none will come after us."[56]

The voyage lasted three years and eighteen days, a staggering achievement: "After escaping innumerable dangers and suffering a long series of hardships . . . it is computed we run over a greater space of sea than any ship ever did before us; since taking all our tracks together, they form more than thrice the circumference of the globe."[57] Just as a glimpse of New York changes one's view of all other cities in the world, Forster's points of reference shifted as a result of his voyage around the world. After his return, he spoke of the North Sea as "the *Great Puddle*."[58] Nevertheless, the circumnavigation, with all its gains in immediate experience, brought to light everything that had yet to be discovered. "Thus it still remains for future navigators, to continue our discoveries in the South Seas, and to take more time in investigating their productions. Several parts of the Pacific Ocean are still untouched by former tracks; for instance, the space between 10° S. and the line across the whole ocean, from America to New Britain; the space between 10° S. and 14° S. included between the meridian of 140° and 160° west; the space included between the parallels of 30° and 20° S. and the meridian of 140° and 175° west; the space between the southernmost of the Friendly Islands, and New Caledonia, and that between New Caledonia and New Holland."[59] There was still a big world out there.

Hardship

"You were on the great voyage?" Kaiser Joseph II asked Georg Forster in August 1784 when Forster was granted an audience with him in Vienna. For Forster and his contemporaries, it was not *a* voyage around the world—any number of which might follow—it was *the* great voyage. Even the fact that Cook had completed three circumnavigations in quick succession could not take away from the singularity of this extraordinary experience. "Did you suffer much?" the Kaiser inquired, demonstrating for a second time not only his interest but also his perceptiveness. Forster informed him he had brought back a "somewhat weak stomach" from the journey.[60] That was an understatement.

Contrary to all romantic notions about sailing since the time when people were reliant on it, a multiyear voyage was above all one thing: an ordeal. Not only because of external dangers, but particularly because of spatial conditions on board. The *Resolution* was a converted coal freighter of some thirty-seven by eleven yards. It lay flat in the water, which was an advantage when approaching foreign shores, and, unlike a warship, it had a large cargo hold, which was indispensable on long journeys. Because it was nearly shipwrecked on the first circumnavigation, Cook brought on board enough fabricated lumber to construct two schooners, one for each ship.

> One should take into consideration how much space such vessels must occupy in a ship, and bear in mind that all the storerooms are crammed with things; on the deck between the mainmast and the foremast are kept five large and small boats; that the sides of the forecastle are covered with huge sheet and bower anchors, as well as other not inconsiderable anchors; that the hold is packed with several hundred casks, some sixty to seventy usually filled with water, the same number with sauerkraut, and far more with salt beef and pork, flour, peas and ship's biscuit, and others full of wine and brandy; that a large quantity of coal is stored in the lowest part of the hold, serving both as ballast to weigh the ship properly down in the water and for daily use in the galley; and that many cables lie between decks, each over a hundred fathoms long, and some the thickness of a man's leg. It is indeed astonishing how in a space that contains four hundred and eighty tons, each ton of which requires forty-four square feet, one hundred and twenty men can find room or, if we grasp this, how with barely digestible food, constant exertion, and the strain of the most severe living conditions, they can remain healthily and in good spirits for three years?[61]

They had cows and bulls, sheep, goats, fowl, and dogs on board, some 27,000 kilograms of rusk (a hard, dry biscuit, or twice-baked bread), almost 20,000 liters of beer, nearly 9,000 kilograms of sauerkraut, more than 47,000 liters of potable water, almost 1,000 liters of olive oil, nearly 8,000 kilograms of flour, 900 kilograms of butter, almost 14,000 kilograms of salted beef, and nearly 13,000 kilograms of salted pork—and so on.[62] Mealtimes aboard the ship were monotonous. For breakfast, the seamen had wheat porridge; in the late morning, they got a portion of rum diluted with water; four days a week, lunch was salted meat, pea soup with boiled-down meat stock, and sauerkraut, which was supposed to prevent scurvy. On the other days the men had to make do with a "hard dumpling made of flour." Dinner was rusk. When potable water wasn't being rationed, every man was permitted as much as he wanted. During their months-long bouts at sea, between ice-bergs and in cold weather, "salt meat, our constant diet, was become loath-some to all, and even to those who had been bred to a nautical life from their tender years: the hour of dinner was hateful to us, for the well known smell of the victuals had no sooner reached our nose than we found it impossible to partake of them with hearty appetite." Sauerkraut was supposed to help the salted meat go down, "without becoming fully aware of the foul, half-rotten taste." No one can understand, Forster concluded, what it means to have "to get by for years at a time on rotten salt meat and mouldy ship's biscuit."[63]

Cramped quarters prevailed on board, and the cabins were stuffy. Al-though Cook attached great importance to hygiene, the conditions were miserable. "Imagine this low and confined space with hammocks strung closely together. At the best of times it receives little fresh air, and dur-ing bad weather it receives almost none, because the main hatch is covered with a tarpaulin stretched over wooden battens. The breathing of more than eighty men not only pollutes this space and causes an unhealthy warmth, but also penetrates the bedding and hammocks, and even the beams and decks of the ship." In temperate latitudes in the South Seas it was sticky, during the course to the South Pole, by contrast, cold and damp: "The decks, and the floors of every cabin were however continually wet."[64]

During the long months at sea, there was hardly any opportunity to be by oneself or to get any exercise. The ship was comparable to a prison in that everyone on board was accorded only the smallest possible range of motion. "Countless times I found myself on the quarter deck, barely twenty-four paces long, together with twelve to fourteen others marching up and down in pairs, so that we all had to turn after twelve or fifteen paces."[65] It was life lived in the most confined space. What a contrast! For the ship was sur-rounded by an infinite expanse of ocean or the icy wasteland of Antarctica. Yet, amid this vastness was a small sailing vessel, crowded with all of the persons who were so essential to a voyage of this sort.

The purpose of Cook's first circumnavigation had been to observe the transit of Venus and—according to his confidential orders—to explore the possibility of a southern continent.[66] The existence of a great land mass in the southern hemisphere could not be settled, which is why Cook sailed southward for months on his second voyage around the world, farther than anyone before, as he noted in his logbook, and as far as it was humanly pos-sible to go,[67] continuing until the Antarctic winter caused him to turn back to the Pacific. "All our officers, who had made several voyages round the world, and experienced a multiplicity of hardships, acknowledged at present, that all their former sufferings were not to be compared to those of the present voyage, and that they had never before so thoroughly loathed a salt diet."[68]

The hardships of the voyage through the Antarctic waters sapped their strength. Forster returns to the subject again and again. It wasn't long before illness occurred. The daily monotony was dreadful: "We were almost perpetually wrapt in thick fogs, beaten with showers of rain, sleet, hail, and snow, the temperature of the air being constantly about the point of congelation in the height of summer; surrounded by innumerable islands of ice against which we daily ran the risk of being shipwrecked and forced to live upon salt provisions, which concurred with the cold and wet to infect the mass of our blood." Although Cook included sauerkraut in the meal plan, "several of our people had now strong symptoms of sea-scurvy, such as bad gums, difficult breathing, livid blotches, eruptions, contracted limbs, and greenish greasy filaments in the urine."[69] Forster reports that he himself got "excruciating pains, livid blotches, rotten gums, and swelled legs" during the voyage. These precarious symptoms "brought me extremely low in a few days, almost before I was aware of the disorder."[70] Over the course of their long stretch at sea, "a general languor and sickly look . . . manifested itself in almost every person's face, which threatened us with more dangerous consequences." Many of the sailors were "at this time afflicted with severe rheumatic pains and colds, and some were suddenly taken with fainting fits, since their unwholesome, juiceless food could not supply the waste of animal spirits." By the time they reached Tahiti in April 1774, Forster had to stay on board because he was "so ill I could scarcely crawl about." When he did venture ashore, "after walking about thirty yards, I was obliged to turn back and sit down, in order to prevent my fainting away."[71] That was the price he had to pay for his immediate experience of the world.

Perils in the Ice

"Nothing appeared more strange to the several navigators in high latitudes, than the first sight of the immense masses of ice which are found floating in the ocean, and I must confess, that though I had read a great many accounts on their nature, figure, formation and magnitude, I was however very much struck by their first appearance," Reinhold Forster writes in his *Observations made during a voyage round the world, on physical geography, natural history,*

and ethic philosophy from the year 1778.[72] Before "the tip of the iceberg" became a common metaphor, and before it became common knowledge that
there was a larger mass of ice underwater, drifting ice floating on top of
the water was a spectacular sight. In contrast to his father, Georg Forster
was eloquent enough to offer insight into the ever expansive and unknown
Antarctic. For his European readership, his depictions of the icy wasteland
proved to be as incredible and strange as his descriptions of Tahiti were
charming. Every syllable here is worth savoring. In *A Voyage round the World*,
compressed scenic snapshots of the expedition are like ice floes from beyond the borders of the known world. Cook's plan was as simple as it was
daring: He wanted to advance as far south as possible to scientifically test
the assumption—deeply embedded in the Western mind—that there was a
great southern continent: Terra Australis Incognita.

The falling temperatures made the occurrence of icebergs probable. In
December 1772, "the great cold preceded the sight of ice floating in the sea,
which we fell in with on the next morning. The first we saw, was a lump of
considerable size, so close to us, that we were obliged to bear away from it;
another of the same magnitude a little more a-head, and a large mass about
two leagues on the weatherbow, which had the appearance of a white head-
land, or a chalk-cliff."[73] It was their first glimpse into a foreign world, entry
to which required acceptance of unheard-of risks.

Georg Forster was impressed. He marveled at the play of color caused by
the sun on the surface of the ice. He was surprised to see that the ice was not
always white, "but often tinged, especially near the surface of the sea, with
a most beautiful sapphirine or rather berylline blue, evidently reflected from
the water; this blue colour sometimes appeared twenty or thirty feet above
the surface, and was there probably owing to some particles of sea-water
which had been dashed against the mass in tempestuous weather, and had
penetrated into its interstices. We could likewise frequently observe in great
islands of ice, different shades or casts of white, lying above each other in
strata of six inches or one foot high." They sailed "through a great quantity
of packed or broken ice, some of which looked dirty or decaying. Islands of
ice still surrounded us, and in the evening, the sun setting just behind one of
them, tinged its edges with gold, and brought upon the whole mass a beautiful suffusion of purple."[74] He had not counted on such unexpected beauty
at the end of the world.

This veritable stained glass included polar light, too, which turned the
sky into a spectacle. As Forster describes, it consisted of

long columns of a clear white light, shooting up from the horizon to the eastward, almost to the zenith, and gradually spreading on the whole southern part of the sky. These columns sometimes were bent sideways at their upper extremity, and thought in most respects similar to the northern lights (*aurora borealis*) of our hemisphere, yet differed from them, in being always of a whitish colour, whereas ours assume various tints, especially those of a fiery, and purple hue. The stars were sometimes hid by, and sometimes faintly to be seen through the substance of these southern lights, (*aurora borealis*), which have hitherto, as far as I can find, escaped the notice of voyagers. The sky was generally clear when they appeared, and the air sharp and cold, the thermometer standing at the freezing point.[75]

The will to contemplate a sight of nature like this from a scientific perspective, but without revealing any aesthetic impression, is unmistakable. This world made of ice was so forbidding that it proved to be surprisingly beautiful.

The course southward was difficult. The crew was not appropriately equipped, by today's standards, for passage into the polar regions. The weather made them increasingly uncomfortable.

The quantity of impenetrable ice to the south did not permit us to advance towards that quarter; therefore, after several fruitless attempts, we stood on to the eastward, along it, frequently making way through great spots covered with broken ice, which answered the description of what the northern navigators call packed ice. Heavy hail showers and frequent falls of snow continually obscured the air, and only gave us the reviving sight of the sun during short intervals. Large islands of ice were hourly seen in all directions around the sloops, so that they were now become as familiar to us as the clouds and the sea.[76]

On January 17, 1773, they "crossed the Antarctic circle, and advanced into the southern frigid zone, which had hitherto remained impenetrable to all navigators."[77]

Navigating the icy sea was risky, since a sailing vessel in motion can only avert but not reverse. The danger became even greater when there was fog. Because the ice beneath the water could not be judged, the wooden ship was continually set on a collision course. "We were encircled by vast masses of ice which emerged from the sea like floating islands and were even more dangerous because their positions could change, and we often sighting them when it was almost too late to steer the ship past. How often were we terrified by being able to hear the waves breaking on the ice, without being able to lay our eyes on the object of our fear." The weather "daily became more sharp, and uncomfortable, and presaged a dreadful winter in these areas; and, lastly, the nights lengthened space, and made our navigation more dangerous than it had hitherto been."[78] Often they "did not see the sun for a fortnight or three weeks," and "the rigging of the ship was adorned with icicles and covered in sheets of ice."[79] The sailors had to handle the frozen ropes without gloves.

Time and again Forster recounts scenes that played out while sailing the icy sea, and how treacherous the ice made the expedition. At some point during their second passage into the southern polar sea in December 1773,

> the weather, which was already foggy, became thicker towards noon, and made our situation, amidst a great number of floating rocks of ice, extremely dangerous. About one o'clock, whilst the people were at dinner, we were alarmed by the sudden appearance of a large island of ice just a head of us. It was absolutely impossible either to wear or tack the ship, on account of its proximity, and our only recourse as to keep as near the wind as possible, and to try to weather the danger. We were in the most dreadful suspense for a few minutes, and though we fortunately succeeded, yet the ship passed within her own length to windward of it. Notwithstanding the constant perils to which our course exposed us in this unexplored ocean, our ship's company were far from being so uneasy as might have been expected; and, as in battle the sight of death becomes familiar and often unaffecting, so here, by daily experiencing such hair-breadth escapes, we passed unconcernedly on, as if the waves, the winds, and rocks ice had not the power to hurt us.[80]

On one such day when they avoided a collision with an iceberg by a hair's breadth, James Cook noted drily in his logbook that they could not count on escaping all other icebergs quite so narrowly.[81]

It is bizarre, a world made of ice, at once fascinating and forbidding. Not intended for human inhabitants, it exceeds all measure of what is familiar.

This journey to the limits of life, through the staggering expanse of Antarctica, transformed the ship into an insignificant point, enclosed by incredible scenery of ossified nature:

> At six in the evening, we counted one hundred and five large masses of ice around us from the deck, the weather continuing very clear, fair, and perfectly calm. Towards noon the next day we were still in the same situation, with a very drunken crew, and from the mast-head observed one hundred and sixty-eight ice islands, some of which were half a mile long, and none less than the hull of the ship. The whole scene looked like the wrecks of a shattered world, or as the poets describe some regions of hell; an idea which struck us the more forcibly, as execrations, oaths, and curses re-echoed about us on all sides.[82]

A wasteland of ice, imposing and forbidding, and of sublime indifference.

To be exposed to these elements was an unexpectedly trenchant experience. The explorers had ventured into a region that humans did not seem allowed to infiltrate. They were at risk of becoming lost in its endless expanse—quite literally, as an incident from December 1772 shows: Reinhold Forster was out in a small dinghy taking measurements with William Wales, the astronomer on board, when both their lives were put in danger. Reinhold Forster did not include this incident in his published travelogue, as it was not scientifically germane. It is telling that his son took up events like this one in his account of the great voyage, to illustrate the "boundary situation" brought about by reckless conduct toward the vagaries of nature.

The scene was eerie: it was snowing heavily that night. Reinhold Forster and Wales were taking advantage of a calm sea by going out in a dinghy to attempt to gauge the temperature of the sea at that depth.

> The fog encreased so much while they were thus engaged, that they entirely lost sight of both the ships. Their situation in a small four-oared boat, on an immense ocean, far from any inhabitable shore, surrounded with ice, and utterly destitute of provisions, was truly terrifying and horrible in its consequences. They rowed about for some time, making vain efforts to be heard, but all was silent about them, and they could not see the length of their boat. They were the more unfortunate, as they had neither mast nor sail, and only two oars. In this dreadful suspense they

determined to lie still, hoping that, provided they preserved their place, the sloops would not drive out of sight, as it was calm.[83]

It was as if they had dropped off the face of the earth. Were they seized by panic? What was going on inside of them we don't know.

At last they heard the jingling of a bell at a distance; this sound was heavenly music to their ears; they immediately rowed towards it, and by continual hailing, were at last answered from the *Adventure*, and hurried on board, overjoyed to have escaped the danger of perishing by slow degrees, through the inclemencies of weather and through famine. Having been on board some time, they fired a gun, and being within hail of the *Resolution*, returned on board of that sloop, to their own damp beds and mouldering cabins, upon which they now set a double value, after so perilous an expedition. The risks to which the voyager is exposed at sea are very numerous, and danger often arises where it is least expected. Neither can we trace the care of Providence.[84]

This is quite different from Reinhold Forster's account, which makes only brief mention of the episode in the ship's log, without lending it any drama. His dry style, which he knew better than to shed in his later publications, did not capitalize on such experiences for the reader. It remains prosy:

The weather grew so foggy whilst we were about making these Experiments on the Sea, that we did not know, where either of the Ships was, though we had seen both a little while before & were between them both, which was scarce ¼ of a mile distance of either. We haled therefore the *Adventure* & they answered & we rowed hereupon towards her; we found every thing well & went away after having haled the *Resolution* & having fired a gun in order to know where she was, & then we put off again, & it cleared so much up, that we just saw, we were close to near her.[85]

It is Georg, not Reinhold, who is the storyteller, the dramatist of events. What might slip into pure reportage by his father or James Cook is brought to life by Forster. It is not a thirst for adventure that guides him. He does not seek the sensational. His experience of nature is far more at home among the very perceptions that sparked portrayal. Georg Forster was tireless in describing his mental rainfall of impressions. Nothing else was possible for him, since nature seemed to safeguard itself from any sober objectification beyond human sentience.

Dangers of this sort might have sapped the crewmembers' nerves, and the rough waters of the polar sea played a part in their exhaustion. Yet, the main burden proved to be the monotony of forging through ice, as well as the perpetual failure to arrive anywhere. Forster does not conceal his weariness by the monotony of impressions. On their second attempt to reach Antarctica in December 1773, it was feared that the voyage would have to be extended when, once again, they encountered no firm land. Forster could make out a "painful despondence" on the faces of the sailors. "The long continuance in these cold climates began now to hang heavily on our crew," he writes.[86]

> A gloomy melancholy air loured on the brows of our shipmates, and a dreadful silence reigned amongst us. . . . It will appear from hence that this voyage was not to be compared to any preceding one, for the multitude of hardships and distresses which attended it. Our predecessors in the South Sea had always navigated within the tropic, or at least in the best parts of the temperate zone; they had almost constantly enjoyed mild easy weather, and sailed in sight of lands, which were never so wretchedly destitute as not to afford them refreshments from time to time. Such a voyage would have been merely a party of pleasure to us; continually entertained with new and often agreeable objects, our mind would have been at ease, our conversation cheerful, our bodies healthy, and our whole situation desirable and happy. Ours was just the reverse of this; our southern cruizes were uniform and tedious in the highest degree; the ice, the fogs, the storms and ruffled surface of the sea formed a disagreeable scene, which was seldom cheered by the revising beams of the sun; the climate was rigorous and our food detestable. In short, we rather vegetated than lived; we withered, and became indifferent to all that animates the soul at other times. We sacrificed our health, our feelings, our enjoyments, to the honour of pursuing a track unattempted before.[87]

It is worth noting that Forster devoted so much of his travelogue to tedium. After all, it could have been interpreted as a lack of loyalty to the British Crown, on whose patronage the expedition depended. Unlike Cook, Forster did not gloss over or sugarcoat the hardships that resulted from attempting to reach the South Pole. He does not merely describe the usual ordeals of a long journey at sea. Rather, almost shockingly, he tells of just how torturous the journey was through the icy "regions of hell."[88] After months in the polar sea, the *Resolution* resembled a phantom ship:

> All the disagreeable circumstances of the sails and rigging shattered to pieces, the vessel rolling gunwale to, and her upper works torn by the violence of the strain. . . . We had the perpetual severities of a rigorous climate to cope with; our seamen and officers were exposed to rain, sleet, hail, and snow; our rigging was constantly encrusted with ice, which cut the hands of those who were obliged to touch it; our provision of fresh water was to be collected in lumps of ice floating on the sea, where the cold, and the sharp saline element alternately numbed, and scarified the sailors' limbs; we were perpetually exposed to the danger of running against huge masses of ice, which filled the immense Southern Ocean: the frequent and sudden appearance of these perils, required an almost continual exertion of the whole crew, to manage the ship with the greatest degree of precision and dispatch.[89]

The men on board had reached the limits of their endurance. They became stricken with a depressive malaise. "We may add to these the dismal gloominess which always prevailed in the southern latitudes, where we had impenetrable fogs lasting for weeks together, and where we rarely saw the cheering face of the sun; a circumstance which alone is sufficient to deject the most undaunted, and to sour the spirits of the most cheerful."[90]

Nevertheless, on January 30, 1774, they reached the southern latitude of 71° and 10 minutes—farther south than any expedition before them. Edmond Halley, in January 1700, had only been able to press as far as 52° south latitude. Now, however, an ice field opened up before them; it extended from east to west, a borderline stretching into the unknown. An icy end to the world, where, as Cook remarked, it was not possible to advance even an inch farther south. They could not have known that the coast of the Antarctic mainland was less than one hundred miles away. Thus, they declared that the existence of a great southern continent had been refuted. Cook did not rule out that there might still be a landmass at the South Pole, but the mythical chimera of a Terra Australis Incognita had been dispelled. Forster, though, insisted on unambiguous terms. For him, this "rendered it evident, that a continent does not exist in the temperate southern zone."[91]

Sunny Arcadia

On August 16, 1773, the two ships were finally approaching Tahiti after months in the ice—only to narrowly avoid catastrophe again. The Tahitians were already paddling out in canoes to meet the *Resolution* and the *Adventure*, when the ships began to drift into the coral reef surrounding the island. It was mainly the *Resolution* that was in danger. "At midday the Resolution bumped into the reef there of coral rocks; she soon, however, refloated," Reinhold Forster wrote, downplaying it in his travelogue. Even Cook saw the incident as scarcely worth mentioning. Although he made no secret of the danger in which they found themselves—"Our situation became more and more dangerous"[92]—dangers were meant to be overcome. Indeed, it was only with extreme effort that they managed to rescue the *Resolution*. Even so, they lost some of the ship's anchors. Regardless of rank, the men worked until exhaustion, towing the *Resolution* away from the reef with a rope winch attached to some marooned boats. Cook was beside himself, cursing and shouting himself hoarse. Then, a breeze picked up, freeing the ship from the dangerous current and releasing it into the open sea.[93]

This scene did not serve the dramaturgy of Forster's narrative, which was intent on describing the paradise of the South Seas after the ordeal of Antarctica and, in doing so, portraying the unspoiled appeal of this new world. As revealed by Reinhold Forster's travel notes, which his son frequently referenced, they first sighted Tahiti on the evening of August 15. The following day they attempted to make landfall; however, because of the difficult conditions, they did not manage it until the 17th. Georg Forster's description of his first glimpse of Tahiti on the evening of August 15 embodies a sense of possibility: "In the evening about sun-set, we plainly saw the mountains of that desirable island, lying before us, half emerging from the gilded clouds on the horizon. Every man on board, except one or two who were not able to walk, hastened eagerly to the forecastle to feast their eyes on an object, of which they were taught to form the highest expectations, both in respect of the abundance of refreshments, and of the kind and generous temper of the natives, whose character has pleased all the navigators who have visited them." In spite of the nautical difficulties, Forster wants the kingdom of the South Seas to take center stage as a refuge for the living: "We resolved to forget our fatigues and the inclemencies of southern climate; the clouds which had hitherto hung lowering upon our brows were dispersed; the loathed images of disease and the terrors of death were fled, and all our cares at rest."[94]

The description of their approach to O-Aitepieha Harbor is tantamount to an enthusiastic crescendo. Forster had already eagerly captured the vegetation from their first stop in New Zealand, which afforded them the greenery they had been longing for after 122 days at sea. He could not get enough of the "wild landscape" of New Zealand: "[It] consists of steep, brown rocks, fringed on the summits with over-hanging shrubs and trees; on the right there is a vast heap of large stones, probably hurried down from the impending mountain's brow, by the force of the torrent."[95] Although Forster tells himself that, after their period of deprivation at sea, "the most barren rock would have been a welcome sight," it was always the forests in their "original, wild, primeval state of nature" that attracted his attention.[96] Tahiti now exceeded all expectations and proved to be of even greater charms.

Forster describes their arrival in Tahiti much as William Hodges, the painter on board the *Resolution*,[97] depicted it—as a still life, in which life appears in the description only toward the end:

It was one of those beautiful mornings which the poets of all nations have attempted to describe, when we saw the isle of O-Taheitee, within two miles before us. The east-wind which had carried us so far, was entirely vanished, and a faint breeze only wafted a delicious perfume from the land, and curled the surface of the sea. The mountains, clothed with forests, rose majestic in various spiry forms, on which we already perceived the light of the rising sun: nearer to the eye a lower range of hills, easier of ascent, appeared, wooded like the former, and coloured with several pleasing hues of green, soberly mixed with autumnal browns. At their foot lay the plain, crowned with its fertile bread-fruit trees, over which rose innumerable palms, the princes of the grove. Here every thing seemed as yet asleep, the morning scarce dawned, and a peaceful shade still rested on the landscape. We discerned however, a number of houses among the trees, and many canoes hauled up along the sandy beaches. About half a mile from the shore a ledge of rocks level with the water, extended parallel to the land, on which the surf broke, leaving a smooth and secure harbour within. The sun beginning to illuminate the plain, its inhabitants arose, and enlivened the scene.[98]

Forster's sketches substantiate his skills as a painter. Here he carries them over into his prose.

Forster creates an idyllic description brimming with charm, like those he first encountered while translating Bougainville's travelogue.[99] Bougainville's description of his arrival in Tahiti had also attempted to capture the magic of his first impression:

> The aspect of this coast, elevated like an amphitheatre, offered us the most enchanting prospect. Notwithstanding the great height of the mountains, none of the rocks has the appearance of barrenness; every path is covered with woods. We hardly believed our eyes, when we saw a peak covered with trees, up to its solitary summit, which rises above the level of the mountains, in the interior parts of the southernmost quarter of the island. Its apparent size seemed to be no more than thirty *toises* in diameter, and grew less in breadth as it rose higher. At a distance it might have been taken for a pyramid of immense height, which the hand of an able sculptor had adorned with garlands and foliage. The less elevated lands are interspersed with meadows and little woods; and all a-long the coast there runs a piece of low and level land, covered with plantations, touching on one side the sea and on the other bordering the mountainous parts of the country. Here we saw the houses of the islanders amidst bananas, cocoa-nut, and other trees loaded with fruit.[100]

In Bougainville's treatment, the landscape is described like an image, painterly in its rendering.

Forster, too, expressly highlighted the relationship between his language and painting. At one point he speaks of a wild landscape "as one of the most beautiful which nature unassisted by art could produce." He mentions only one painter by name in *A Voyage round the World*: Salvator Rosa, a seventeenth-century painter famous for his landscapes. Forster's use of words is comparable to Rosa's use of colors to achieve emotional effects. He writes of a spot that is "one of the most beautiful places that I had ever seen, and could not fail of bringing to remembrance the most fanciful description of poets." The connection between poetry and painting has a long tradition. Horace wrote in *Ars poetica* that a poem is like a painting: "ut pictura poesis." It is in this tradition that Forster, ever the observer in his sketching and writing, can speak of "word paintings."[101]

Forster's descriptions of nature and of the Tahitian landscape are lovely through and through. They seem to be the realization of all his longing, a paradise, a new Arcadia. Forster later left instructions for his gravestone to be inscribed, "I too have lived in Arcadia—*Et in Arcadia ego.*" For the European reader, Tahiti became "at the same time a geographically real and mythically ideal place," as Klaus H. Börner notes[102]—an idealization that had an enduring influence on the imaginations of the literati back home.

As that first day came to an end, nighttime epitomized tranquility beneath skies of the South Seas: "All night the moon shone clear in a cloudless sky, and silvered over the polished surface of the sea, while the landscape lay before us like the gay production of a fertile and elegant fancy."[103] Bathed in early morning light, the next day was redolent with life's promise:

> We contemplated the scenery before us early the next morning, when its beauties were most engaging. The harbor in which we lay was very small, and as smooth as the finest mirrour, and the sea broke with a snowy foam around us upon the outer reef. The plain at the foot of the hills was very narrow in this place, but always conveyed the pleasing ideas of fertility, plenty, and happiness. Just over against us it ran up between the hills into a long narrow valley, rich in plantations, interspersed with the houses of the natives. The slopes of the hills, covered with woods, crossed each other on both sides, variously tinted according to their distances; and beyond them, over the cleft of the valley, we saw the interior mountains shattered into various peaks and spires, among which was one remarkable pinnacle, whose summit was frightfully bent to one side, and seemed to threaten its downfall every moment. The serenity of the sky, the genial warmth of the air, and the beauty of the landscape, united to exhilarate our spirits.[104]

For Forster, Tahiti is an unparalleled experience. "I have learned to appreciate nature above all else since I came to know her in Taheiti and have sought her out in so many places in Europe to no avail," Forster writes upon his return.[105] Tahiti is not just another impression from his voyage; it is the euphoric culmination of his immediate experience of nature. Forster's concept of what is natural has been newly defined, having found in Tahiti a botanical and topographical antecedent for what philosophers like John Locke or Jean-Jacques Rousseau had conceived merely as an illustration of the natural condition of humanity. Nature here, or so it promises at first glance, provides the setting from which humankind may emerge in its natural state, where human gentleness corresponds to natural loveliness.

The result of these [reflections] was a conviction, that this island is indeed one of the happiest spots on the globe. The rocks of New Zeeland appeared at first in a favourable light to our eyes, long tired with the constant view of the sea, and ice, and sky; but time served to undeceive us, and gave us daily cause of dislike, till we formed a just conception of that rude chaotic country. But O-Taheitte, which had presented a pleasing prospect at a distance, and displayed its beauty as we approached, became more enchanting to us at every excursion which we made on its plains. Our long run out of sight of land might have been supposed at first to have had the same effect as at New Zeeland; but our stay confirmed instead of destroying the emotions which we had felt at the first sight.[106]

First and Final Sightings

Among the privileges of immediate experience afforded by a voyage around the world in the eighteenth century is sighting something for the first time. Hitherto unknown glimpses of foreign plants, animals, and people became accessible to European observers. The magic of discovering the world was followed by the gradual disillusionment that each new sighting jeopardized its existence. Forster could not yet anticipate the fragility of the natural habitat; however, he likely recognized how easily the balance in human societies could be upset. This proved to be significant to his political thinking.

Nature was still a kingdom of staggering abundance to him. When he explored the forests of New Zealand in March of 1773, he was profoundly impressed to find "forests [that] have never been touched by human industry, but have remained in the rude unimproved state of nature since their first existence." Forests such as these were scarcely known in Europe. "Our excursions into them gave us sufficient grounds for this supposition; for not only the climbing plants and shrubs obstructed our passage, but likewise numbers of rotten trees lay in our way, felled by winds and old age. A new generation of young trees, of parasitic plants, ferns, and mosses sprouted out of the rich mould to which this old timber was reduced by length of time, and a deceitful bark sometimes still covered the interior rotten substance, whereon if we attempted to step, we sunk in to the waist." He came upon birds that "had not yet undergone any changes from the hands of mankind. . . . Numbers of small birds which dwelt in the woods were so little acquainted with men, that they familiarly hopped upon the nearest branches, nay on the ends of our fowling-pieces, and perhaps looked at us as new objects, with a curiosity similar to our own." Yet impressions of this kind were precious few, the radius of immediate experience being rather small. Goethe marveled at the "luxuriant vegetation" in Naples during his Italian journey.[107] He sighted flora in the Mediterranean plant kingdom, such as palms and agave, which had not yet found their way north at that time.[108] The impression of tropical forests, still untouched by the extraction of raw materials, was indescribably exotic by comparison and remained hidden from European travelers.

On the "botanical rambles" that Forster and his companions undertook to gather "the treasures of nature in countries hitherto unknown,"[109] they regularly came across animals and plants that were not part of the coordinate system of the known world: "On our part, we perceived a new store of animal and vegetable bodies, and among them hardly any that were perfectly similar to the known species, and several not analogous even to the known genera."[110] Indeed, the reward for the hardship wrought by their journey is knowledge of the world, an original, immediate experience of the landscapes, animals, plants, entire regions, and human communities. The youthful vigor of Forster's travelogue can be attributed to his unclouded view of the world—the threat to which he cannot fully sense yet.

A few decades later, the vulnerability of societal and ecological systems became undeniable. Adelbert von Chamisso, who, between 1815 and 1818, took part in the Russian expedition to explore a navigable passage from Europe to the Pacific, already recognized that their incursion would introduce change to unknown peoples. "These customs, which I still could see," he writes woefully of his impressions of the Hawaiians, "are no longer carried out on these islands, and the language of the liturgy is destined to die away. No one appears to have thought of investigating and thus saving from oblivion that which could contribute to our understanding of the externals of the law of this people, perhaps shed light upon its history, and perhaps the history of mankind."[111] These are already images of the past, as gains in immediate experience irrevocably violate the undisturbed presence of that which is experienced.

Even Alfred Russel Wallace, who traveled the Malay Archipelago between 1854 and 1862 and who devised a theory of evolution around the same time as Darwin,[112] was profoundly aware that from the history of discovery there emerged a history of loss. When he caught sight of a king bird of paradise deep in the Malay jungle for the first time, he was enraptured by this "thing of beauty," and he thought of "the long ages of the past, during which the successive generations of this little creature had run their course—year by year being born, and living and dying amid these dark and gloomy woods, with no intelligent eye to gaze upon their loveliness; to all appearance such a wanton waste of beauty. Such ideas excite a feeling of melancholy."[113] Wallace had a fine feeling for the fragility of the natural habitat:

> It seems sad, that on the one hand such exquisite creatures should live out their lives and exhibit their charms only in these wild inhospitable regions, doomed for ages yet to come to hopeless barbarism; while on the other hand, should civilized men ever reach these distant lands, and bring moral, intellectual, and physical light into the recesses of their virgin forests, we may be sure that he will so disturb the nicely-balanced relations of organic and inorganic nature as to cause the disappearance, and finally the extinction, of these very beings whose wonderful structure and beauty he alone is fitted to appreciate and enjoy.[114]

This historical difference between Forster and the soon-to-be theorist of evolution is not insignificant. After all, the sight of nature's beauty, hitherto undiscovered and withheld from humankind for so long, leads Wallace to the insight "that all living things were *not* made for man"—because the richness and abundance of life are independent of the beholder. "The cycle of their existence has gone on independently of his, and is disturbed or broken by every advance in man's intellectual development." Charles Darwin poses the same question during his circumnavigation from 1831 to 1836, upon sighting rare animals: they "play so insignificant a part in the great scheme of nature, one is apt to wonder why they were created."[115]

Considerations such as these do not yet play a role for the young Forster. And even in later years he declared the question unanswerable, "insofar as the climate of each place, as originating cause, can contribute to the existence of certain organic bodies with their peculiar forms and properties." Forster casts his wistful enthusiasm for man's connectedness with nature in traditional forms of expression, when, for example, he says "that the study of nature always delivers new fragments to God's knowledge." When he notices enormous quantities of plankton in the ocean, as if "the whole ocean seemed to be in a blaze," Forster writes, "There was a singularity, and a grandeur in the display of this phaenomenon, which could not fail of giving occupation to the mind, and striking it with a reverential awe, due to Omnipotence. The ocean covered to a great extent, with myriads of animalcules; these little beings, organized alive, endowed with locomotive power, a quality of shining whenever they please, of illuminating every body with which they come in contact, and of laying aside their luminous appearance at pleasure: all these ideas crouded upon us, and bade us admire the Creator, even in his minutest works."[116]

This can certainly be taken for sanctimoniously enthused rhetoric. However, it is evidence of Forster's desire to determine how an emphatic concept of nature can clearly and centrally integrate humanity into the whole of the world. Human knowledge and the natural world should still be compatible. "Nothing is as irresistible as truth, as nature," Forster says. And even that which is hidden in nature awaits humanity to behold it. It is in this vein that Forster refers to "those magnificent flowers whose abundance and delicateness surpass all, which flaunt themselves on the stalk of the flare-thistle for just one hour at night only to wither before dawn. Nature could not confer for any duration on such delicately immersed life, and so she tosses it into the barren wilderness, sufficient in and of itself, to wither unnoticed, until a human being, as I understand the word, that rarest being in creation, finds and enjoys its fleeting phenomenon!"[117]

Over the course of his life, Forster abandoned the Christian interpretation of nature as revealing "the true ways of divine providence in the wonderful works of creation" and the understanding that nature was a "divine creator in the distribution of its goods." He later expressly replaced the idea of the "highest providence of God" (*summa Dei providentia*) with the "highest wisdom of nature" (*summa naturae providentia*).[118] Although the abundance of nature may be staggering, it is still custom-tailored to human needs. In one of his most charming descriptions, Forster portrays tropical nature as a cornucopia, kept at the ready for humanity. In his 1784 essay "Der Brodbaum" (The breadfruit tree), he depicts the tropical sun as preparing

the finest and most spiritual mixtures of saps; instead of common rubber and resin, camphor and benzoin flow from the trees' wounds, or elusive fragrant oils permeate the bark, filling the blossoms and fruit, and forming that rare spice, for whose possession the European peoples have waged bloody wars. . . . Here the mango and the mangosteen surpass the delicious fruits of all other parts of the world; and spoiled palates that seek fulfillment from constantly new delights can be provided complete satisfaction with more than fifty precious varieties of fruit. This region is also home to numerous beautiful flowers, which captivate more than one sense at the same time. Trees of inner structure and incomparable hardiness tower above the forests; and the noble genus of palms is native in all its species here. These princes of the plant kingdom crane their slender trunks marvelously above all other trees, the feathery treetops spreading out in perpetual green, and there they stand, the inimitable ideal of majestic simplicity. . . . The fish in the seas, the butterflies, and other insects vie for the prize of rarity in form or color. Just as rich is the plumage of countless species of birds. Yet, shimmering above them all is the bird of paradise, like the rarely sighted occupants of an Asian harem, showered with many-coloured gold, and dipped in the purple of dawn. Finally, enter the larger animals, in all their manifold development, but with one creature at the top [here he means the orangutan] in whose human-resembling form nature perhaps wanted to show how precisely she could emulate creation's masterpiece, at least outwardly, with her designs.[119]

Even in its intoxicating splendor, nature still holds humanity dear. Its abundance may exceed all of our expectations, but we are still the "masterpiece."

Yet Forster perceives nature as a menacing force, too. Humanity does not impinge on the fragile natural habitat so much as the human cultural habitat is destroyed by nature. Thus, he sees "with a shudder . . . the terrible disorder which nature . . . can unloose in the works of man. The pressure of rocks piled atop each other, while underground, carelessly hollowed out molehills, too, can collapse, bringing death and destruction, leaving no trace of man's former industriousness behind." The implication here is a repeal of classical anthropocentrism, in which the world is created for humans. Forster even addresses it explicitly at one point in a letter when he speaks of "man being a fool, when he says the world is made for his sake."[120] Finding shelter is a tenuous human endeavor in the face of nature's superiority. Anyone who sails around the world for three years in a converted coal freighter knows this.

It is all the more significant that Forster not only devoted himself to the magic of his early glimpses of human cultures but also became conscious — as Chamisso later did — at least to some extent, of the balance of civilized systems that could be all too easily destroyed. The discovery of "simple child[ren] of nature" did not leave those cultures unchanged.[121] Each point of contact affected their way of life. With this fragility in mind, Forster developed what had been missing from his perspective on the immutability of nature: an emerging consciousness of the precarious balance of existing societies. He could already sense that first sightings always had the potential to be final sightings, too. Although the voyages of discovery transpired in the radiance of the Enlightenment, and not the dusk of declining worlds, the discoverer's transforming influence was already coming into view for Forster.

Attempts by the English to cultivate the cultures they visited—agriculturally, for example, through the creation of vegetable gardens and the introduction of pigs—were unsuccessful. Forster notes the gradual shift in moral norms among indigenous populations through their contact with the Europeans, remarking on the "irretrievable harm" inflicted on indigenous populations by the explorers' "corrupting their morals." "If these evils were in some measure compensated by the introduction of some real benefit in these countries, or by the abolition of some other immoral customs among their inhabitants, we might at least comfort ourselves, that what they lost on one hand, they gained on the other; but I fear that hitherto our intercourse has been wholly disadvantageous to the nations of the South Seas; and that those communities have been the least injured, who have always kept aloof from us, and whose jealous disposition did not suffer our sailors to become too familiar among them."[122] The history of exploration is a history of tragedy: "If the knowledge of a few individuals can only be acquired at such a price as the happiness of nations, it were better for the discoverers, and the discovered, that the South Sea had still remained unknown to Europe and its restless inhabitants." Years later, in Lichtenberg's 1782 *Göttingen Pocket Calendar*, Forster expressed the hope that the great distance between the South Sea Islands and Europe would provide protection for their societies, since the new expeditions did "not spread trade, but rather merely science."[123] At the same time he notes that it is solely "the curiosity and self-interest of the Europeans" that causes them to set out for distant lands, only to emerge as an "odious phenomenon" there.[124]

This remark's criticism of civilization nevertheless contains a point on Forster's philosophy of history: unlike nature, which he still thought of as immutable, societies are variable. Societies can be set in motion. This can be either lamented—or accelerated.

Noble Savage?

Rousseau's conception of a "natural man" (*homme naturel*) arose in response to the spoiled civilization of his day, providing a speculative basis for his critique of society. Nothing "can be more gentle than man in his primitive state."[125] Nevertheless, the cliché of the "noble savage" in his primitive state as the ideal of pure gentleness is not evident in Rousseau's text. For Rousseau, man in his primitive state suffers the conversion from a peaceable self-love (*amour de soi*), intended for self-preservation, into a competitive love of self (*amour propre*). The naturalness of this first man is not pure; rather, contained within that naturalness is the germ of development, which Rousseau diagnoses as the history of the malady of modern civilization. Since it remains unsettled whether Forster was familiar with Rousseau's work while he was writing *A Voyage round the World*,[126] though, philological accuracy plays a subordinate role here. A Rousseau lecture may have been the source for the concept of the noble savage, or it may have found its way to Forster as a mere phrase. Regardless, we can assume its historical presence was such that Forster would have grappled with it.[127]

Forster, however, draws a different picture of the indigenous people he encountered on the voyage. His views on that which was foreign to him unmistakably change during the course of his travels. Little by little, as his travelogue can attest, he sheds his Eurocentric perspective.[128] On Madeira, the first stop on their voyage, a European point of view still clings to his descriptions: The common people are

> of a tawny colour, and well shaped, though they have large feet, owing perhaps to the efforts they are obliged to make in climbing the craggy paths of this mountainous country. Their faces are oblong, their eyes dark; their black hair naturally falls in ringlets, and begins to crisp in some individuals, which may perhaps be owing to intermarriages with negroes; in general they are hard featured, but not disagreeable. Their women are too frequently ill-favoured, and want the florid complexion, which, when united to a pleasing assemblage of regular features, gives our Northern fair ones the superiority over all their sex.[129]

Forster does not shy away from letting his aesthetic criteria influence his judgment of foreign peoples. He describes the inhabitants of the island of St. Jago—today, Santiago, Cape Verde—as being of "a middle stature, ugly, and almost perfectly black with frizzled wooly hair and thick lips, like the most ill-looking kind of negroes."[130]

Although he regards the inhabitants of "this torrid zone," by which he means black South Africans, as being naturally inclined to "sloth and laziness," he tests out a cultural-philosophical approach, in which he consults their social situation to explain this characteristic and to correct for his moralizing judgment. If they are resigned to their idleness, it is because of the despotism of their rulers, and they must "become indifferent to improvement, when they know the attempt would only make their situation more irksome."[131]

His description of New Zealand's Maori vacillates between approval and disgust. Forster says, "Their colour was of a clear brown, between the olive and mahogany hues, their hair jetty black, the faces round, the nose and lips rather thick but not flat, their black eyes sometimes lively and not without expression; the whole upper part of their figure was not disproportionate and their assemblage of features not absolutely forbidding." However, there was also "a certain stench which announced them even at a distance, and [an] abundance of vermin which not only infested their hair, but also crawled on their clothes, and which they occasionally cracked between their teeth."[132]

Forster not only describes what he finds compelling and repelling about a people but also compares different peoples. Particularly stark is the contrast between the people of Tierra del Fuego and those of Tahiti. The people of Tierra del Fuego presented him with an exhibit of misfortune. They seemed "without the smallest degree of curiosity" and were

> short, not exceeding five feet six inches at most, their heads large, the face broad, the cheek-bones very prominent, and the nose very flat. They had little brown eyes, without life; their hair was black and lank, hanging about their heads in disorder, and besmeared with train-oil. On the chin they had a few straggling short hairs instead of a beard, and from their nose there was a constant discharge of *mucus* into their ugly open mouth. The whole assemblage of their features formed the most loathsome picture of misery and wretchedness to which human nature can possibly be reduced.[133]

The "wretched outcasts" exhibited to him "the strangest compound of stupidity, indifference, and inactivity." This group represents to Forster the lowest state of civilization. Their "mode of life approaches nearer to that of brutes, than that of any other nation. It is indeed very probable, that they are the miserable out-casts of some neighbouring tribe, which enjoys a more comfortable life; and that being reduced to live in this dreary inhospitable part of Tierra del Fuego, they have gradually lost every idea, but those which their most urgent wants give rise to."[134]

The Tahitians, by comparison, prove to be of antique grace, beauty, and charm. The foreignness of tattooing—still unknown in Europe at the time—does not prevent Forster from likening Tahitian body art and dress to that seen in Greek statues:

> If this dress had not entirely that perfect form, so justly admired in the draperies of the ancient Greek statues, it was however infinitely superior to our expectations, and much more advantageous to the human figure, than any modern fashion we had hitherto seen. Both sexes were adorned, or rather disfigured, by those singular black stains, occasioned by puncturing the skin, and rubbing a black contour into the wounds, which are mentioned by former voyagers. They were particularly visible on the loins of the common men, who were almost naked, and exhibited proof how little the ideas of ornaments of different nations agree, and yet how generally they all have adopted such aids to their personal perfection.[135]

The erotic freedom on Tahiti and its neighboring islands leaves Forster speechless by how "the simple child of nature, who inhabits these islands, gives free course to all his feelings, and glories in his affection towards the fellow-creature."[136]

Beyond the level of detail in his ethnological descriptions, which cannot be fully cataloged here, it is interesting to note Forster's intentions for them and what he omits from them. Although he does not withhold his immediate impressions of these foreign peoples—"participant observation" comes to mind[137]—he also does not propose, as Jörn Garber emphasizes, "a historical system of human progress," into which it would be tempting to force his findings.[138] To be sure, a tiered model of civilizations was so ubiquitous during the Enlightenment that Forster would not have been free of it completely.[139] He even fell back on this schema in his 1787 essay *Cook, the Discoverer*, and he developed it into a four-tiered model in his 1789 essay "Leitfaden zu einer künftigen Geschichte der Menschheit" (Guide to a future history of humanity). Tiered models of this kind commonly came to serve as the basis for cultural imperialist racism. The handshake between colonialism and missionary work meant that the exploitation and reeducation of seemingly backward peoples were grounded in a historical-philosophical foundation. All the more striking is Forster's attempt to use the tiered model against Christoph Meiners' crude race theory, according to which there are two unrelated types of people: those who are physically and morally perfect;

and those who are ugly and depraved.[140] Forster's use of a tiered model in his analysis serves the exact purpose of allocating a tier of education in human history even to the "savages," thus preserving them from any racist compartmentalization. In "Antwort an die Göttingischen Recensenten" (1778), Forster states—in response to critical objections to his travelogue—his goal of demanding "a halt to unjust prejudice, in which we in Europe dare to grant ourselves virtue, and the savages nothing but iniquity and evil nature." Moreover, in his essay "Der Brodbaum," written in the summer of 1784, he describes black slaves as "the most ill-treated beings," defending them from the prejudice that they are merely "bastards from apes and men" and granting them "consciousness and reason" and thus unconditional humanity.[141]

Increasingly, though, this civilizing hierarchy's ideology of progress was being placed into question by Forster's ethnological perspective at the time he wrote *A Voyage round the World*. He succeeded in doing much more, according to Garber, "by using a comparative method to collect the advantages and disadvantages of the savage and civilized societies." The comparison was not always favorable to Europeans. Forster is able to counter the immorality of the "savages" as being observable in individual cases: "Vicious characters are to be met with in all societies of men; but for one villain in these isles, we can show at least fifty in England, or any civilized country."[142] Only immediate experience can teach what humanity is. Only an accrued diversity of experiences can outline its shape. Humanity, one must unequivocally say, is not a European phenomenon. Forster's descriptions, in their differentiation, take aim at what Garber calls a "geography of world cultures."[143]

Forster is well aware of the creative aspects of narrative description. "It is necessary to be acquainted with the observer, before any use could be made of his observations." And the caveat applies to all ethnological descriptions: he and his father rendered their impressions "each . . . according to his sensations."[144]

Rousseau's speculative theory of a "primitive state" was itself a thoroughly European idea. At the sight of suffering by the "Pecherais," the people of Tierra del Fuego, Forster rejects once and for all the premise that natural humanity had an original state, instead emphasizing the achievements of the civilized world: "If ever the pre-eminence of a civilized life over that of the savage could have been reasonably disputed, we might, from the bare contemplation of these miserable people, draw the most striking conclusions in favour of our superior happiness." Philosophers, who have only dreamed of a primitive state, have "either had no opportunity of contemplating human nature under all its modifications, or . . . have not felt what they have seen."[145] The blessed "primitive state" is the brainchild of the armchair theorist.

Although Forster may have employed the concept of a natural state of humanity from time to time, his references to those he found antipathetic have the aim of breaking with the ideal of a primitive state. There is a significant clue to his renunciation of the European idealization of exotic foreign experiences. William Hodges accompanied Cook on his second circumnavigation, making etchings for his travelogue. Forster did not like his etchings. He accused Hodges of not conveying "any adequate idea" of the inhabitants of the Tonga Archipelago that he depicted. It was as if "Hodges [had] lost the sketches and drawings which he made from nature in the course of the voyage, and supplied the deficiency in this case, from his own elegant ideas." The problem with such portrayals of their travels was that they exhibited "pleasing forms of antique figures and draperies, instead of those Indians of which we wished to form some idea."[146]

To demonstrate his realism, Forster did not shrink from offering descriptions that ran contrary to the wishful hope of Europeans that there were paradise-like conditions in the South Seas. While visiting the Friendly Islands, he writes of lepers,

> In some of them the disorder had risen to a high degree of virulence; one man in particular had his whole back and shoulders covered with a large cancerous ulcer, which was perfectly livid within, and of a bright yellow all round the edge. A woman was likewise unfortunate enough to have all her face destroyed by it in the most shocking manner; there was only a hole left in the place of her nose; her cheeks were swelled up and continually oozing out a purulent matter; and her eyes seemed ready to fall out of her head, being bloody and sore. These were some of the most miserable objects I recollect ever to have seen.[147]

Descriptions of this sort did not fit with the desired image of the noble savage. The conditions in Pacific Arcadia ought to have been paradisiacal. Forster reports sightings that—in the words of Schiller—were as unsettling as "when the plague rages among angels."[148]

Among Maneaters

The Age of Enlightenment discovered the human race in novel ways. The Christian faith had ensured the unity of humanity through its descent from Adam and Eve. This connectedness now came to be founded on new terms. The cultural differences between peoples might have been just as great, and the exotic peoples on the edges of the known world might have seemed just as foreign; yet the faculty of reason proved to be the connecting element.

There were severe tests, though. Bloodshed and mayhem loomed everywhere, bellicose confrontations and vindictiveness, hatred and discord. In Europe, rumors circulated about cases of cannibalism in the distant regions of the world. It seemed so egregious that Immanuel Kant deemed it an exaggeration. The Portuguese had reported maneaters in Africa, who purposely fattened their victims. "But we ought not to give credence to fables of this type so easily," Kant argues, "as experience has shown that these people only slaughter prisoners of war who have been captured alive, and even then only with the greatest ceremony." As for cannibals, Kant continues: "In accordance with human nature, there cannot be many, or, more likely, perhaps none at all."[149] Still, so it would seem, humanity was immune to falling apart through grievous self-annihilation in the taboo form of self-consumption.

However, there were reports of cannibalism that rose to the respectability of lexical information. Johann Heinrich Zedler's 1739 *Universal Lexicon aller Wissenschafften und Künste* unsettled its reader with this entry: "Man-eaters, meaning, on the whole, cannibals, Hottentots, and other wild Indians; it is sufficiently known from virtually all descriptions of travel occurring to their regions how they, as well as those foreigners who become lost among them, as well as their enemies who become captured in battle, are hacked to pieces in the most abominable manner, and are thereupon either boiled or roasted until they can be consumed off the bone." But those "wild Indians" were far enough away that they did not disturb any European intellectual games about the universality of humanity. Cannibals belonged to the realm of mythical fabrication, where, for example, one could read Marco Polo's account of people who possessed "heads like dogs and teeth and eyes likewise" and were said to be "man-eaters."[150]

In October 1773 the *Resolution* and the *Adventure* landed in New Zealand. It was their second time there that year. Forster did not like the New Zealanders. His first impression, from March 1773, in spite of good faith, had already been one of revulsion. The "state of barbarism in which the New

Zeelanders may justly be said to live" made them liable in his mind, "more than any other nation to resolve upon the destruction of their fellow-citizen, as soon as an opportunity offered."[151] It was only going to get worse.

When James Cook, Reinhold Forster, and William Wales went on shore, they came upon human entrails in a pile on the beach. "They were hardly recovered from their first surprize, when the natives shewed them several limbs of the body, and expressed by words and gestures that they had eaten the rest. The head of the lower jaw-bone, was one of the parts which remained, and from which it plainly appeared, that the deceased was a youth about fifteen or sixteen years old. The skull was fractured near one of the temples, as it seemed by the stroke of a pattoo-pattoo," a wooden club. Some days later they found a canoe on the beach, with "a carved head ornamented with bunches of brown feathers, and a double-forked prong projected from it, on which the heart of the slain enemy was transfixed." Questioning the natives revealed that the young man had been a member of a warring group. The rest of his body they had consumed. Suddenly, rumors about cannibalism were proving to be true, as they circulated in the educated milieus of the European Enlightenment, and ceased to be merely tall tales. "Now we have with our own eyes seen the inhabitants devouring human flesh, all controversy on that point must be at an end," Forster determines, not without horror. The New Zealanders were "man-eaters"![152]

Further proof was required to dispel the notion that cannibalism was inconceivable, and it resulted in an anthropological experiment on board the *Resolution*. An Englishman had in his possession the head of a young man, which he had exchanged for a nail and brought on board the ship, where he displayed it from the balustrade. When the New Zealanders saw it, "they possessed an ardent desire of possessing it, signifying by the most intelligible gestures that it was delicious to the taste." So the Englishman cut a piece of flesh from the cheek and gave it to them. When they refused to eat it raw, it was broiled over the fire, "after which they devoured it before our eyes with the greatest avidity." When Cook returned, they "repeated the experiment once more in his presence."[153]

Forster might now be expected to judge the barbaric New Zealanders. But the true experiment was just beginning, because the sight of the New Zealanders eating a young man's flesh

operated very strangely and differently on the beholders. Some there were who, in spite of the abhorrence which our education inspires against the eating of human flesh, did not seem greatly disinclined to feast with them, and valued themselves on the brilliancy of their wit, while they compared their battle to a hunting-match. On the contrary, others were so unreasonably incensed against the perpetrators of this action, that they declared they could be well pleased to shoot them all. . . . A few others suffered the same effects as from a dose of ipecacuanha. The rest lamented this action as a brutal depravation of human nature, agreeably to the principles which they had imbibed.[154]

It was not anticipated that some Europeans would reveal themselves to be just as inclined to do the same to the savages!

This can be considered a key scene for an anthropology that forms its assumptions about people from immediate experience. For Forster, all speculation about human nature, all explication of humanity that is guided by reason, is settled in one fell swoop. "Philosophers who have only contemplated mankind in their closets, have strenuously maintained, that all the assertions of authors, ancient and modern, of the existence of men-eaters are not to be credited," but they now had proof of the contrary before their eyes. What humanity is consequently becomes accessible though observation and experience alone. What one might call Forster's naturalist anthropology is built on an epistemology that is guided by experience and oriented toward immediate experience.[155]

As it is so guided and so oriented, however, it is held up to the centralism of universal reason, which means that humans have measures of worth that are independent of experience. "The action of eating human flesh, whatever our education may teach us to the contrary, is certainly neither unnatural nor criminal in itself." Forster develops a cultural relativism on his voyage around the world, which seeks to refuse the dominance of European thinking. Nevertheless, he interprets his comrades' impulse to shoot the natives as a reaction to the incident of cannibalism, "in order to punish the imaginary crime of a people whom they had no right to condemn."[156]

Forster cites a case of cannibalism in Germany in 1772. Initially triggered by famine, it became a practice for the cannibal to murder his victims "as a very delicious food." Against this backdrop, the difference between cultures is not unbridgeable, which is to say, we all have the potential to be cannibals. The overwhelming indignation with which the Europeans spurn the practice does not make them better people in Forster's eyes. "But though we are too much polished to be cannibals, we do not find it unnaturally and savagely cruel to take the field, and to cut one another's throats by thousands, without a single motive, besides the ambition of a prince, or the caprice of his mistress! Is it not from prejudice that we are disgusted with the idea of eating a dead man, when we feel no remorse in depriving him of life?" Michel de Montaigne states in his *Essays*, with regard to the mariners' reports of cannibalism among the savages, that Europeans "surpass them in every kind of barbarism." In times of war and torture, there is "more barbarity to eating a man alive than in eating him dead."[157]

How can cannibalism be explained, though? Was Kant correct in speculating that it was the result of bellicose confrontations? Forster searches for an explanation along these lines. It is known that "revenge has always been a strong passion among barbarians, who are less subject to the sway of reason than civilized people, and has stimulated them to a degree of madness which is capable to all kinds of excesses. The people who first consumed the body of their enemies, seem to have been bent upon exterminating their very inanimate remains, from an excess of passion."[158] Finding the flesh wholesome and palatable, he continues, it became a habit to eat the enemies they killed. In this way, the strangest thing is by no means incomprehensible. The unity of the human race remains intact. There is no insurmountable rift through cultures. The unifying element is only affect, not reason.

Forster's defense of the unity of the human race, as well as his refusal to regard cannibals as monsters, was grimly tested during a third stop in New Zealand in October 1774. On January 8, 1773, the *Resolution* and the *Adventure* had "entirely lost sight of" each other in the Antarctic ice because of "exceedingly thick fog."[159] However, they had prepared for such an eventuality, and in May of that year, they were able to reunite during a preordained stop in Queen Charlotte Sound, New Zealand. On October 29, 1773, the two ships became separated once and for all during a storm off the coast of New Zealand. A full year later now, the *Resolution* was to land in New Zealand for the third and last time, when the mariners heard from the natives that "an European vessel had put into the harbor some time ago; but that in a quarrel with the inhabitants, all her people had been killed and eaten." Horror spread among the men, since their experience had shown the occurrence of cannibalism to be a reality, plucked from the realm of myth and fantasy. Forster reported, "[This news] alarmed us greatly, as we apprehended that this vessel was most probably the Adventure."[160] They questioned the natives but were reassured that it must have been a misunderstanding. Only upon their return to England did they learn what had happened. The *Ad-*

venture had sailed into Queen Charlotte Sound, New Zealand, in November 1773, to make preparations for its return to England. Ten men under the direction of a Mr. Rowe set out for a final leave onshore to gather "scurvygrass." Two days passed, and they had not returned. An armed search party was dispatched, which met with fleeing New Zealanders. Finally, one of the seamen found a deserted canoe, inside which he discovered "several mangled limbs of their comrades and some of their cloaths." The mere sight of the natives sparked images of them "probably dressing human flesh. Horror chilled the sailors' blood in their veins, but the next moment they glowed with the fierce ardour of revenge, and cooler reason was obliged to give way to the powerful impulse. They fired and killed several of the natives." When they looked further ashore, "they found many other limbs of their friends packed into baskets, and particularly a hand, which they knew to be that of the unfortunate Rowe," who had led the dispatched search party just days before.[161] They could not prepare themselves for what they saw: "The dogs of the New Zeelanders were meanwhile eating the entrails strewn on shore."[162]

When cultural anthropology is based on observation, its explanatory power is dependent on the distance of the observer from the phenomenon being observed. Forster does not join in the frenzy of outrage. Rowe had "combined with many liberal sentiments the prejudices of a naval education, which induced him to look upon all the natives of the South Sea with contempt, and to assume that kind of right over them with which the Spaniards, in more barbarous ages, disposed of the lives of the American Indians." The Englishman, so it can be read, was basically a barbarian, incapable of encountering foreign cultures without resentment. Thus, when conflict broke out onshore, the seamen reacted excessively. One of the natives stole a sailor's jacket, whereupon the English "immediately began to fire, and continued to do so till all their ammunition was spent; . . . the natives had taken this opportunity to rush upon the Europeans and had killed every one of them."[163]

The fate of the slain is so unthinkable that Forster defers to a basic tenet of anthropology, namely, that humanity's affects are stronger and more determinant than reason. Consequently, Forster places the true blame on the Englishmen. Although the theft of the jacket was wrong, the "rash action of revenging this theft with death, and most probably revenging it indiscriminately on a whole body of natives, must have provoked them to retaliate." From an anthropological and cultural philosophical perspective, their actions were understandable: "Born to live our stated time on this globe, every one who puts a premature period to our existence here, offends the laws of the Creator. The passions are wisely implanted in our breast for our preservation; and revenge, in particular, guards us against the encroachments of others. Savages do not give up the right of retaliating injuries; but civilized societies confer on certain individuals the power and the duty to revenge their wrongs."[164] The brutality by the savages is presented as an anthropologically justified reflex to the injustice done by the Europeans, whose retaliation for the theft was unwarranted. A lack of centralized violence gives rise to autonomous action, which, in civilized societies, is reserved for institutions.

Forster views cannibalism as an integral part of a universal cultural history. In his later essay "Über Leckereyen" (On delicacies), which appeared in Lichtenberg's 1788 *Göttingen Pocket Calender*, Forster describes the phenomenon of "man-eating" as arising from a "very natural instinctual desire." In doing so, he integrates the phenomenon of anthropophagy, the eating of human flesh, into the history of humanity. Cannibalism is no longer an occurrence at the margins of humanity, practiced by peoples who have not yet left behind their barbaric status. Even European cultural history, Forster surmises, has known cannibalism. There may be no historical witnesses, except to the practice of human sacrifice as an ersatz form. "For it is known that with all nations this kind of barbarism turned into the custom to sacrifice humans, and that this religious ceremony persisted for a long time, even with an improvement in culture and morals. Thus the Greeks, Carthaginians, and Romans still sacrificed human beings to their gods when their culture had already reached its zenith."[165] For Forster, cannibals are not creatures beyond the brink of humanity, but rather members of the human community who are to be recognized. Even in light of the terrible events of Cook's second circumnavigation, Forster can still locate in the most disconcerting conduct an anthropological basis and a civilizing step forward, which, when considering the cultural differences, serve to render comprehensible what initially seemed completely inconceivable.

Bloodshed and Mayhem

A blood-red thread is woven into the colorful narrative tapestry of *A Voyage round the World*: the story of violence, mayhem, and murder. Accounts of brutality interrupt the flow of storytelling, drawing attention to the knots that thicken the narrative thread. In his later look back, *Cook, the Discoverer,* Forster is determined to extol the exploration of the world as human progress, but *A Voyage round the World* paints a different picture—perhaps because the impressions were still fresh, the destruction so immediate. The acts of violence during his three years at sea not only were left unembellished by Forster; when told in chronological order, they take on an intensity.[166]

As these records reveal, at the time of the voyage around the world, Forster had not yet become so hardened that the senseless use of force failed to shock him. In August 1772, a few weeks after first setting sail, they were just leaving Madeira when a stray swallow, drenched by rain, came aboard the *Resolution.* "I dried it, and when it was recovered, let it fly about in the steerage, where, far from repining at its confinement, it immediately began to feed upon the flies, which were numerous there." They let the swallow go, but it returned, "being sensible that we intended it no harm." It stayed on board a while longer but then could not be found. For Forster, his presumption amounted to the first sin of the voyage: "It is more than probable that it came into the birth of some unfeeling person, who caught it in order to provide a meal for a favourite cat." He acknowledges that "in the long solitary hours of an uniform navigation, every little circumstance becomes interesting to the passenger; it is therefore not to be wondered at, if a subject so trifling in itself as putting to death a harmless bird, should affect a heart not yet buffeted into sensibility."[167] At the first intimations, he speaks of murder.

Forster views the sailors on board as constituting "a body of uncivilized men, rough, passionate, revengeful, but likewise brave, sincere, and true to each other." They are a source of violence toward each other, because "by force of habit even killing is become so much their passion, that we have seen many instances during our voyage, where they have expressed a horrid eagerness to fire upon the natives on the slightest pretences."[168] The threshold to violence was low. Forster might have shed his Eurocentrism—or at least tried not to let it pervade—but for the rest of the men on board, this was not the case. When a few Tahitians sought to trade coconuts, breadfruit, and the like, for small items such as coral, knives, and nails on board the ship, they were accused of cheating for having bartered the same wares more than once: "The thieves were turned out of the vessel, and punished with a whip, which they bore very patiently."[169]

The violence quickly escalated, because of "how easy it is to provoke the mariner to sport with the lives of Indians." On the Friendly Islands, a native pilfered a coat from the ship and was shot as he absconded. He nevertheless managed to make it to shore and into a crowd of natives, where the English shot at him again. "By [this] means several innocent people were wounded."[170] The next day, another native stole from one of their cabins "mathematical books, a sword, a ruler, and a number of trifles of which he could never make the least use." He was hunted, weapons were fired at him, and he was pursued into the sea. "At last one of our people darted the boat-hook at him, and catching him under the ribs, dragged him into the boat."[171] Although bleeding badly, he managed to escape again. What became of him, we do not learn from Forster. The incident shows Forster how "the harmless disposition of these good people could not secure them against those misfortunes, which are too often attendant upon all voyages of discovery."[172]

Property crimes, above all, were strictly avenged by the English. There is a template in the history of ideas for this: John Locke outlines in his 1690 *Two Treatises of Government* a model of how property is lawfully gained. Each person has property on his own person, as well as that which he has appropriated through physical labor and the work of his hands. Although the world belongs to all, it remains free to the individual who acquires property through labor. The amassing of perishable goods beyond the measure of possible consumption is wrong, according to Locke, since it deprives the community of what could have been used by others. The moral limit to amassing goods lapses, however, as soon as a nonperishable good, namely money, takes the place of perishable goods. *In nuce*, what we have here is the modern theory of legitimated capital gains, which set into motion the ideal of a static estate-based society. Anyone who works a lot is permitted to possess a lot, too, and in doing so, to leave behind the social class he was born into. Counter to the Christian proviso that it is easier for a camel to go through the eye of a needle than for a rich man to enter the kingdom of God, Locke provides a legitimation for adopting the morally dubious nature of prosperity through work. A strong conception of property was the consequence, which the state then took into account through the protection of private property.

An eloquent example of this understanding of property is Forster's 1779 account "Das Leben Dr. Wilhelm Dodds" (The life of Dr. Wilhelm Dodd). Forster describes the trial of the London clergyman William Dodd, which was closely followed by the public. Dodd was not exactly unknown, even by the German-speaking public. Goethe claims in *Dichtung und Wahrheit* (Poetry and truth) that it was through Dodd's anthology *Beauties of Shake-*

speare that he was first introduced to the English poet. Dodd had committed forgery, was caught, and although he was offered indemnity by those he had swindled, and despite impressive public support—twenty-three thousand signatures were collected in a petition to pardon the beloved parson—he was sentenced to death and publicly hanged. The king alone could have issued a pardon, but he was counseled by his advisers that "in a country propelled to action, the security of property, as one of the most essential assets, could not be protected strictly enough; that (especially in England) this security belongs to the foundation of the constitution." Property crimes, even minor ones, shake the foundation of the state! Forster makes a point of mentioning that on June 27, 1777, "a young man," perhaps still a child, by the name of Joseph Harris was to be put to death alongside Dodd because of a "committed street robbery."[173] This nearly emphatic eighteenth-century conception of property, taken together with the willingness to protect possessions by draconian punishments, is important to note, as it forms the backdrop for James Cook's conduct, too. When an indigenous person on the island of Tanna tries to cheat Cook out of what was owed him in a trade, Cook shot him unceremoniously: "A musket, charged with small shot was fired into his face."[174]

Cook was intent on letting encounters with the indigenous proceed peaceably—for reasons having to do with procurement of provisions. The Royal Society had issued instructions on Cook's first circumnavigation, prescribing restraint when dealing with the indigenous and strictly controlling the use of firearms to avoid bloodshed.[175] Cook would have understood these instructions to be binding for his subsequent two voyages, too.

Nevertheless, armed conflict occurred again and again. In June 1774 Cook, Forster, and a few other crew members were attacked with spears on one of the Friendly Islands in the South Pacific. Cook's musket misfired. Reports Forster, "The natives threw two spears: captain Cook narrowly escaped one of them by stooping; the other slid along my thigh, marking my cloaths with the black colour with which it was daubed. We tried to fire again, and at last my piece, loaded with small shot, went off." What happened to the man Forster fired upon? Did he kill him? Or merely injure him? Forster remains silent on the subject, but John Elliott, a lieutenant on board the *Resolution*, later spoke of it. Forster only wounded the attacker, but his act of self-defense undoubtedly saved Cook's life.[176]

An incident on Tanna in August 1774 ultimately placed the moral integrity of the entire expedition into question for Forster. He had undertaken the voyage around the world with the aim of "representing mankind in a favourable light" and fostering cultural understanding. But he explains: "We had now passed a fortnight amidst a people who received us with the strongest symptoms of distrust, and who prepared to repel every hostile act with vigour. Our cool deliberate conduct, our moderation, and the constant uniformity in all our proceedings, had conquered their jealous fears. They, who in all probability had never dealt with such a set of inoffensive, peaceable, and yet not despicable men; they who had been used to see in every stranger a base and treacherous enemy, now learnt from us to think more nobly of their fellow creatures." This comment may be interpreted as idealization; yet, Forster's narrative structure relies on escalation. Scarcely had the hearts of the Tanna opened with "a new disinterested sentiment, of more than earthly mould, even friendship," writes Forster, than the paradise of human communality collapsed.[177] There on the beach in Tanna he found a distressing scene: "We beheld two natives seated on the grass, holding one of their brethren dead in their arms. They pointed to a wound in his side, which had been made by a musket-ball, and with a most affecting look they told us 'he is killed.'" In their language, writes Forster, "they express this more strikingly by one word, *markom*." Forster depicts the scene poignantly—comparable to an *Imago pietatis*, a devotional image of the sufferings of Christ.[178]

The dead Tanna man had literally crossed a line that had been drawn in the sand by an Englishman and was shoved back by a sentry. Accustomed to such treatment, he "refused to be controuled on his own island by a stranger; he prepared once more to cross the area, perhaps with no other motive at present than that of asserting his liberty of walking where he pleased." The sentry pushed him back once more, this time with greater force. The rebuffed man laid an arrow on his bow and took aim at the sentry, which the English soldier classified as an attack and "leveled his musket and shot him dead."[179]

Forster's sympathies are entirely with the man who was shot dead. It was "murder." The "injurious treatment" caused the Indian to defend his rights, while the blame fell to the cruel and treacherous people "who so grievously violated the laws of hospitality." Forster does not endeavor in the least to defend his countrymen. The incident on the beach in Tanna disgusts him: "Thus one dark and detestable action effaced all the hopes with which I had flattered myself. The natives, instead of looking upon us in a more favourable light than upon other strangers, had reason to detest us

much more, as we came to destroy under the specious mask of friendship."
It could not have been worse. On Tanna, "some amongst us lamented that
instead of making amends at this place for the many rash acts which we had
perpetrated at almost every island in our course, we had wantonly made it
the scene of the greatest cruelty."[180]

Jan Philipp Reemtsma might find it "patently absurd,"[181] but there were
regularly acts of violence, and it is only by chance that no fatalities occurred
until the end of their stay on Tanna. Indeed, Forster arranges the facts according to the needs of the story, and his criticism of violent deeds by the
English is as free of a correct chronology as Rousseau's criticism of society
on the premise of a "primitive state" was free of fact. Forster arrives at this
point alone: The English are assassins!

This notably isolated incident of excessive violence places into question the overarching morality of such expeditions of discovery and, accordingly, the hope for civilization by the western world. "It is unhappy enough
that the unavoidable consequence of all our voyages of discovery, has always
been the loss of a number of innocent lives."[182] A different balance was
struck in Forster's later essay *Cook, the Discoverer*. However, the "tristes
tropiques" are unmistakably inscribed in *A Voyage round the World*.

A Community of Equals

One of the most surprising things Forster experienced in Tahiti was the
tenuously developed hierarchy among the indigenous people. He was familiar with the conditions in European courts, where everything was at the
discretion of the powerful. There, one had to "wait, hope, fawn," without
getting anything in the end. In the court of the Russian empress, he had to
endure "how much of the Russian state itself and the amenability of societal dealings are determined by station, which everyone has there and has
been assigned according to the class of people in which they were born,"
such that "each person who holds rank over another lords over him, too, to
some extent."[183] Nor would he forget the indignity his father went through
to receive payment after their strenuous journey to the Volga on behalf of
the empress. In retrospect, the months spent in the Antarctic ice seemed
like a backdrop against which the lush vegetation of the Pacific islands stood
out all the more spectacularly, just as the encrusted power structures of the
aristocracy throw into stark relief the conditions in Tahiti.

The Tahitian social structure was not foreign to him; in fact, it resembled that of the Europeans: "Under one general sovereign, the people are distinguished into the classes of aree, manahoùna, and towtow, which bear some distant relation to those of the feudal systems of Europe." However, while the three-class society led to insurmountable divisions in the European courts, in Tahiti Forster perceives it to be a mere organizing principle that lacks steep gradients between the classes: "The evident distinction of ranks which subsists at Taheitee, does not so materially affect the felicity of the nation, as we might have supposed. . . . The simplicity of their whole life contributes to soften these distinctions, and to reduce them to a level."[184] Simplicity is the price society must pay for the near equality of its members.

This slight differentiation in standing, which corresponds to various tasks, does not serve to disadvantage the island's inhabitants. In the life of the Tahitians, Forster sees "the picture of real happiness."[185] He argues that their living conditions facilitate a carefree life. The breadfruit tree, on which Forster wrote a separate essay in 1784, serves as a symbol of the "beneficent product of nature." Its fruit effortlessly secures the livelihood of the Tahitians: "A person can amply live off the yield of three breadfruit trees for eight months."[186]

For a natural anthropology based on immediate experience, the sight of an elderly Tahitian man provides undeniable evidence of the inducement to an effortless lifestyle. "His head, which was truly venerable, was well furnished with fine locks of a silvery grey, and a thick beard as white as snow descended to his breast. His eyes were lively, and health sat on his full cheeks. His wrinkles, which characterize age with us, were few and not deep; for cares, trouble, and disappointment, which untimely furrow our brows, cannot be supposed to exist in this happy nation." The naturalness of this people's way of life, it would seem to Forster, is comparable to Rousseau's speculation about a primitive state without civilizing pathologies. The starkly ritualized dealings with one another in the European courts is thought to be the hallmark of a civilized people, whereas it is the ostensible lack of taboos in the Tahitians' way of life that impresses Forster: "Where the means of subsistence are so easy, and the wants of the people so few, it is natural that the great purpose of human life, that of multiplying the number of rational beings, is not loaded with that multitude of miseries which are attendant upon the married state in civilized countries. The impulses of nature are therefore followed without restraint, and the consequence is a great population, in proportion to the small part of the island which is cultivated." Forster believes he has an unspoiled community before his eyes. His view of the Tahitians is shaped by the ideal "that philanthropy seems to be natural to mankind, and that the savage ideas of distrust, malevolence, and revenge, are only the consequences of a gradual depravation of manners."[187]

In this community of equals, Forster notes, the king is accorded a role that does not compel respect through distance. Except that all subjects have to bare their shoulders in the presence of the king, Forster observes no ritual of power. In the "familiarity between the sovereign and the subject," Forster detects that "perhaps the origin of their government was patriarchal," because "the lowest man in the nation speaks as freely with his king as with his equal, and has the pleasure of seeing him as often as he likes. The intercourse would become difficult as soon as despotism should begin to gain ground. The king at times amuses himself with the occupations of his subjects, and not yet depraved by the false notions of an empty state, often paddles his own canoe, without thinking such an employment derogatory to his dignity." During an audience of some of the sailors with the king, he spoke "with great affability to our common people" on crew, and all the people "exerted themselves in acts of hospitality and testimonies of friendship from the lowest subject to the queen." Without the trappings of power, social distinctions do not make for an unbridgeable distance. The rulers themselves remain one with the people. "The chief who had visited us on board and accompanied us to the shore, was in nothing different from the common people, not even in his dress; it was only from the obedience which was paid to his orders that we concluded his quality."[188]

This idea paves the way to viewing those in power as functionaries who serve for a set term. They lack that theological dignity with which the European monarchs guarantee their aloofness. On the second visit to New Zealand in November 1773, Forster was reacquainted with Teiratuh, a former commander who now presented himself as no longer holding that title. The formerly impressive speaker "seemed to be degraded to a simple fishmonger. It was with some difficulty that we recognized his features under this disguise, upon which he was taken into the cabin, and presented with some nails." Forster thus learned the concept of governance *on call*, rather than as idealistically conceived. He is awestruck. For him, the equality of all people is not a demand made by political reason, but rather the result of immediate experience. "At O-Taheitee there is not, in general, that disparity between the highest and the meaneast man, which subsists in England between a reputable tradesman and a labourer."[189]

In his contemplation of Tahitian society, Forster has an early immedi-

ate experience of what Jürgen Habermas describes as the "horizontal association of citizens."[190] A society is possible in which there are distinctions
in function but the dominant organizing principle is equality for all. Before
the French Revolution, an idea like this must have seemed outrageous and
ostensibly far removed from reality. It proved to be all the more decisive in
Forster's political thinking after his experiences in the South Seas.

The idea of equality for all people, as consistent with a horizontal society, was taking shape at this time in Europe, too. Jean-Jacques Rousseau,
an intellectual pioneer of the Great Revolution, in his 1758 *Letter to d'Alembert*, describes a scene from his childhood that later became the nucleus for
his idea of an egalitarian society. In one long passage, Rousseau recounts
how, after completing a military exercise, the soldiers began to dance with
the officers:

> The Regiment of Saint-Gervais had done its exercises, and, according
> to the custom, they had supped by companies; most of those who formed
> them gathered after Supper in the St. Gervais square and started danc-
> ing all together, officers and soldiers, around the fountain, to the basin
> of which the Drummers, the Fifers and the torch bearers had mounted.
> A dance of men, cheered by a long meal, would seem to present noth-
> ing very interesting to see; however, the harmony of five or six hundred
> men in uniform, holding one another by the hand and forming a long rib-
> bon which wound around, serpent-like, in cadence and without confusion,
> with countless turns and returns, countless sorts of figured evolutions, the
> excellence of the tunes which animated them, the sound of the Drums,
> the glare of the torches, a certain military pomp in the midst of pleasure,
> all this created a very lively sensation that could not be experienced coldly.
> It was late; the women were in bed; all of them got up. Soon the windows
> were full of Female Spectators who gave a new zeal to the actors; they
> could not long confine themselves to their windows and they came down;
> the wives came to their husbands, the servants brought wine; even the
> children, awakened by the noise, ran half-clothed amidst their Fathers
> and Mothers. The dance was suspended; now there were only embraces,
> laughs, healths, and caresses. There resulted from all this a general emo-
> tion that I could not describe but which, in universal gaiety, is quite natu-
> rally felt in the midst of all that is dear to us. My father, embracing me, was
> seized with trembling which I think I still feel and share. "Jean-Jacques,"

he said to me, "love your country. Do you see these good Genevans? They are all friends, they are all brothers; joy and concord reign in their midst."[191]

Heinrich von Kleist also imagined a classless society. In his story *The Chilean Earthquake*, which appeared in the *Morgenblatt für gebildete Stände* in 1807, he describes the situation after a catastrophic earthquake: "In the fields, as far as the eye could see, people of all the classes were lying without distinction, princes and beggars, patrician wives and peasant women, the state's officials and day labourers, monks and nuns, they were seen extending pity and help to one another and gladly sharing whatever they might have saved for their own sustenance, as if the general misfortune had made one family of all who had survived."[192]

Rousseau and Kleist clearly sought to locate the equality of people not in an idea but in an experience—factually contrived or fictionally described. The idea of a perfect society remains abstract and cold compared to the warm experience of a community. While an egalitarian society is presented here merely as an environmental nucleus for the burgeoning idea, Forster is talking about having already experienced a completely realized and at least tendentially classless society. Government is not a utopia for Forster, but something that rests on the equality of all. His acceptance of equality for all people, as well as the possibility of a societal form that takes this equality into account, rests on true experience. This is the difference between Forster and those European poets and thinkers who idealize equality and fraternity based on their selective experiences. It is this difference that accounts for their varying degrees of willingness to be radicalized for the revolution. While the visionaries must rely on the implementation of an ideal, Forster can reclaim for himself his desire to bring about what he already knows from his own immediate experience.

For Forster, enforcing the parity of all citizens means defending the unity of the human race. In this regard, Tahitian society is in no way inferior in its simplicity to European society. It is just different—without being completely different in nature. For Forster, as for Rousseau, the simplicity of societal conditions points to a quality of being undefiled by civilization. Goethe asked Forster, at their first encounter in September 1779, about the "Southerners," whose "simplicity pleased him." Simplicity is not grounds for disqualification. Thus, every imperialist gesture is prohibitive for Forster—until it becomes "possible for Europeans to have humanity enough to acknowledge the indigenous tribes of the South Sea as their brethren."[193] This was written a good two decades before the French Revolution.

The Aggrieved Rights of Mankind

The great prospect that Forster holds out in his *Voyage round the World* is the promise of humanity's unity with nature. Communality will arise from the naturalness of mankind, Forster hopes, with the knowledge to avoid the excesses of civilization in the encrusted states of central Europe.

> Those who are capable of being delighted with the beauties of nature, which deck the globe for the gratification of man, may conceive the pleasure which is derived from every little object, trifling in itself, but important in the moment when the heart is expanded, and when a kind of blissful trance opens a higher and purer sphere of enjoyment. Then we behold with rapture the dark colour of lands fresh prepared for culture, the uniform verdure of meadows, the various tints upon the foliage of different trees, and the infinite varieties in the abundance, form, and size of the leaves. Here these varieties appeared in all their perfection, and the different exposure of the trees to the sun added to the magnificence of the view. Some reflected a thousand dancing beams, whilst others formed a broad mass of shadow, in contrast with the surrounding world of light. The numerous smokes which ascended from every grove on the hill, revived the pleasing impressions of domestic life; nay my thoughts naturally turned upon friendship and national felicity, when I beheld large fields of plantanes all round me, which, loaded with golden clusters of fruit, seemed to be justly chosen the emblems of peace and affluence. The cheerful voice of the laboring husbandman resounded very opportunely to complete this idea.[194]

A sentimental idyll, certainly. But scenes like this are written for European readers, who are to be shown the possibility of leading a life that is freed from the day-to-day and no longer forced by the corset of societal norms.

Flashes of envisioning humanity as a large family recur in Forster's sprawling narrative. It may come across as naive to those who still have in the back of their minds the brutality of the history of conquest since Columbus, along with the imperial gap between the "civilized" and the "savages."

Forster writes of the Tanna Islanders: "They shared the abundant produce of their soil with their new acquaintance, being no longer apprehensive that they would take it by force. They permitted us to visit them in their shady recesses, and we sat down in their domestic circles with that harmony which befits members of one great family. In a few days they began to feel a pleasure in our conversation, and a new disinterested sentiment, of more than earthly mould, even friendship, filled their heart." A few years later, in 1785, Friedrich Schiller, in his "Ode to Joy," echoed Forster's experiences of an amicably connected community of equals on the shores of the South Pacific. "Beggars and princes will become brothers" is how he puts it[195]—and it would be naive not to perceive the political undertones in Forster's idyll.

Yet the paradisiacal society in Tahiti, as Forster depicts it, already contains the seeds of decline. It does not merely serve as an auspicious contrast for European societies; it also offers a glimpse into the dynamics that can disrupt a fragile balance of equals. "How long such an happy equality may last, is uncertain," he wonders. And he sees their "absolute idleness" as being the germ of their demise.[196] It is their decadence that fortifies the distance between otherwise scarcely detectable class differences.

Forster writes of the "Taheitian drone" to illustrate the start of this demise.[197] While out on a walk, he comes upon a house, "where a very fat man, who seemed to be a chief of the district, was lolling on his wooden pillow." Two servants were preparing him dessert. "While this was doing, a woman who sat down near him, crammed down his throat by handfuls the remains of a large baked fish, and several breadfruit, which he swallowed with a voracious appetite. His countenance was the picture of phlegmatic insensibility, and seemed to witness that all his thoughts centred in the care of his paunch."[198] An encounter which could have been dismissed as an isolated case Forster extrapolates into cultural-anthropological pessimism:

> We had flattered ourselves with the pleasing fancy of having found at least one little spot of the world, where a whole nation, without being lawless barbarians, aimed at a certain frugal equality in their way of living, and whose hours of enjoyment were justly proportioned to those of labour and rest. Our disappointment was therefore very great, when we saw a luxurious individual spending his life in the most sluggish inactivity, and without one benefit to society, like the privileged parasites of more civilized climates, fattening on the superfluous produce of the soil, of which he robbed the laboring multitude.[199]

What Forster could not have known was that the man was being subjected to a harsh Polynesian cultural taboo, which did not permit him to touch food with his hands.[200]

In interpreting the scene as decadent, Forster's ethnological description once again contains political undertones that anticipate, like a roll of thunder, the diagnosis of grievances in European class society. What might be the case in Tahiti has implications for Europe, too. Of course, a good decade passed before Forster fleshed out its application in *Cook, the Discoverer*. When reminded of the "Tahitian drone," he writes: "The fattened idler, who exists at the expense of the working and subservient class in O-Taheiti and Europe alike, is nothing but a miscarriage of government."[201]

Yet, on the shores of Tahiti, Forster is already venturing an outrageous thought: If people are free by nature and created equal, class differences contradict how he sees the fundamental rights of man beginning to develop there. This must have consequences: "At last the common people will perceive these grievances, and the causes which produced them; and a proper sense of the aggrieved rights of mankind awaking in them, will bring on a revolution." Like a cognitive earthquake, he unexpectedly drops the term that anticipated the political consequences of his immediate experience in the South Seas. Because revolution is part of "the natural circle of human affairs."[202] This remark looked like sheet lightning for the events that were to come.

III

Interludes

1776–1788

:: :: :: :: ::

After his return to England in 1776, Forster visits Paris, where he meets Georges-Louis Leclerc de Buffon, the renowned author of *Histoire Naturelle*, and the American writer, scientist, and inventor Benjamin Franklin. In 1778 Forster moves to Germany and begins his friendship with Friedrich Heinrich Jacobi. In 1779 Forster accepts a professorship in natural history in Kassel, marking the start of his friendship with Samuel Thomas Soemmerring and Georg Christoph Lichtenberg. From 1780 to 1785, Forster coedits with Lichtenberg the *Göttingisches Magazin der Wissenschaften und Litteratur*. In 1784 Forster accepts an invitation to chair the natural sciences department at the Polish university in Vilna. His path to Vilna takes him through Leipzig, Dresden, Prague, Vienna—where he is granted an audience with Kaiser Joseph II—Krakow, and Warsaw. In 1785 Forster marries Therese Heyne, daughter of the philologist Christian Gottlob Heyne of Göttingen. They have four children: Therese ("Röschen") was born in 1786, and Clara ("Claire") was born in 1789; their other two children, Luise and Georg, died just a few months after they were born, in 1791 and 1792, respectively. In 1786 Forster engages in a public debate with Immanuel Kant on the concept of race. In 1787 Forster receives an offer from the Russian government to undertake a four-year research trip, whereupon he and his family leave

Vilna. Preparations for the expedition are thwarted by the outbreak of the Russian-Turkish War. In 1788 Forster accepts a position as a librarian at the University of Mainz. Forster's marriage is strained by an affair between his wife and Ludwig Ferdinand Huber.

:: :: :: :: ::

Blue Devils

In the eighteenth century, a voyage around the world was extraordinary. It turned every voyager into a celebrity. Anyone who had been through the ordeals of circumnavigating the globe as an explorer was a welcome guest at all of Europe's receptions. The academies of science opened their gates, and well-compensated professorships became a prospect. Cosmopolitan naturalists were courted, and universities vied in their bids to be graced by a Forster. So one might think, but the reality was different.

The two Forsters received four thousand pounds for the voyage. All expenses were deducted from that sum: travel gear, provisions for their family members remaining in London, and additional expenses incurred on board. The remaining amount was not collected until their return. The four thousand pounds were no longer considered reimbursement for expenses, but rather wages, which is to say, no gratuity from the royal family could be expected. Forster offered King George III his intricately drawn color illustrations for a considerable sum, but the king did not wish to put up the funds to acquire them—a slight to Forster. Debt drove Reinhold Forster to sell off the majority of his son's superb drawings, which included illustrations of rare animals and plants, to Sir Joseph Banks for less than fair value. As a result, 572 pictures disappeared from public view for quite some time. In 1780 Goethe gave a bundle of Forster's drawings to the Duke of Gotha.[1] The two explorers' financial situations were precarious upon their return: "In short, we are on our own again," Forster wrote in September 1776.[2]

Years of financial woes followed, with unremitting work on books, essays, and translations. The translations in particular were arduous work, and Forster groused to his publisher that he was not "a translation machine." His father was continually breathing down his neck; he made use of Georg at his discretion and regarded him as an assistant, fully at his disposal. His domineering ways brought his son to declare resentfully, "No one has ever been so used by his father as I have." In 1778 Forster accepted a professorship in Kassel, but the salary was poor. He lived "in a little room on the third floor." He bitterly remarked, "I remember that I am Buffon's translator. . . . I've corresponded with Earls and written an abecedarian book of natural history; I've sailed around the world only to end up in Kassel teaching twelve-year-old halfwits how to spell in their mother tongue."[3]

In 1784 he accepted an offer from the Polish university in Vilna, where the professorship was better paid, but socially he found himself in a "barbaric land."[4] Forster's journal entries from his journey to Vilna reveal what his letters can only hint at: "God! God! Is this the place that I, a blind and ill-fated man, believed could and must wrest me from all hardship? O my Lord and my God, how am I so utterly unfortunate?—Ill, culpable, unable to pay for anything, penniless, without hope of receiving any money, no furniture, and indeed, most desperately in need; completely idle and suffering here, without books, in an insalubriously cold room . . . ; without company, without the means to even seek such out for myself, without a wagon, a captive in my own hollow." He must admit to not being able to capitalize on his world travels, because "essentially it is of no consequence that one has been around the world; one's value is no greater merely because of it."[5]

Yet, Forster's journey to Vilna, which took him through Leipzig, Dresden, Prague, Vienna, Krakow, and Warsaw, was also a triumph. He was a celebrity, warmly received and passed around everywhere. In Vienna he was granted an audience with Kaiser Joseph II. This only made the social isolation in Vilna harder to take.

The miserable circumstances ensured one thing at any rate: Forster returned to Europe not as a revolutionary but as a slave to his circumstances. The constant scourge of having to provide, in spite of poor health, for his own livelihood, as well as for his father, his mother and his siblings, and, since his marriage in 1785, for his wife Therese (née Heyne) and their two children Therese and Clara—all of this caused politics writ large to recede into the background. Stuck, as Schiller wrote in *The Robbers*, between "insect cares" and "giant projects," between "God-like plans and mouse-like occupations," Forster was at risk of wearing himself out. In addition to teaching, he was writing a botanical thesis, which was published in 1786. His state of mind vacillated between despair and cautious optimism, depending on whether "the weather was good or bad" in his head.[6] He admits to having read Goethe's *The Sorrows of Young Werther* "two, three times in a row" and finding great relief in "a many-houred howl of tears."[7]

In contrast to the grind of work that he was forced back into, the simple life of the South Seas seemed like a piercing memory to him, making his reentry into European society difficult. He had long outgrown the fusty rituals of courtly society and could scarcely resign himself to its hierarchical schemes. The paradisiacal conditions on Tahiti solidified into a painful memory: "Of the thorns in my heart, which I came to know and was not permitted to inhabit in the land of promise, the promised land, where milk and honey flow, where I came to learn what the human race has to worship, and of which I must be deprived evermore—I do not even want to think of it!"[8] Like a metaphor for his fleeting time in Tahiti, Forster's "complete herbarium from the South Sea as has never before been seen and perhaps in this century will not be collected again" was damaged in shipping from London to Germany, along with a cache of handbooks on natural history. When the storage boxes came ashore, everything inside had "utterly decayed into rubbish."[9]

Forster felt out of his element. "I wish to be of use to the world—what can I do? These thoughts have circulated in my head for twelve months," he writes his father in September 1783. Even the professorships in Kassel and Vilna do not seem to have changed his self-doubt. Indeed, Forster recognized early on that he was "not suited to teaching."[10] The lectures caused him "unspeakable toils." Every word of them had to be "read off the page"; otherwise he would not get through them. He suffered from a "form of shyness" that drove him "mad."[11]

He was also haunted by a melancholic temperament; he spoke of "blue devils."[12] He drifts "on the vast sea of dread," loathsome to himself, "sullen, gloomy, melancholic, shedding tears, inconsolable." Time and again it flares up: he is "so unhappy!" His "soul mourns."[13] Cannily, he establishes a psychosomatic context of physical and mental suffering, writing in a letter of 1792, "The more my body is weakened by illness, the more this fatal mood prevails. They are causes of mental pathology."[14]

The "far-flung desert of Vilna," a "dilapidated city with many bleak houses and piles of rubble," became a nightmare for him, epitomizing the way his life was being passed by. But he could not leave Vilna, because he had committed himself to eight years there, and he would have had to pay back the money he had received to cover his travel to Vilna—he was "in debt up to his ears." The foreignness of Vilna remained a determining factor. Despite his "skill speaking and writing all kinds of languages," he could not acquire Polish, which for him was a "severely barbaric language." To make matter worse, the lingua franca of academic life was Latin, which he did not have the time to learn. The formal education he had missed out on was taking its bitter revenge. The "unfamiliar work causes calluses,"[15] he said, because he had to "poke [his] nose in Cicero" to be able to "eke out" his lectures in Latin.[16]

Overall, he lamented the "fatal embarrassment" of his situation in a place that offered him no social happiness.[17] He and his wife settled into their new home as well as could be expected. Their first daughter, Therese, was born on August 10, 1786. However, Forster's familial happiness could not mitigate his feeling that he was condemned to being marginalized. Already on the journey to Vilna he had suffered from loneliness, from "withdrawn-ness," as he called it: "Even among people, in company large and small, among indifferent persons and those who take an interest in me, I am absent, distracted, isolated."[18] During this period, he was at risk of becoming a stranger to himself: "O how little do I know myself still?" All was gone, "beauty, blossom of youth, fire in the eyes—all vanished. Well! That is the price for having voyaged around the world—for that one must go into exile in Vilna!"[19]

Forster never recovered from the torments of the voyage. He paid the price for having circumnavigated the world for the rest of his life. His body remembered the torture that he had been subjected to. His letters and journal entries from later years provide insight about this, as well as his "melancholia or hypochondria." He was often "miserably ill once again," his stomach "ruined by salt fish and spoiled rusk." It was not uncommon for his head to "ache horribly" and "desolately." He could barely run, became fatigued, and was tormented by "pains in [his] legs"; "both knees hurt and the right calf." His legs were swollen, his foot "quite cramped." He lamented his "very tired bones" and his "rheumatism in [his] hands and feet." His fingers would hurt so much that for days he could not "open a door or hold the quill without pain." His suffering, "soon in [his] eyes, soon in [his] stomach, soon in [his] back," when coupled with the "most severe cramp-like swelling in [his] bowels," verged on martyrdom. His stomach was "all a mess."[20] He suffered from colic, "swollen glands," and "chest pains."[21] He got "epidemic river fever" and "horrid pains in his fingers," causing him to lose his fingernails. "And as if this weren't bad enough," he added, "the sharpness descended to my toes to such an extent that for six weeks I was confined to my room and at times could scarcely drag myself from my bed to my desk."[22] His was a daily monotony of pain: "Always the familiar headache, the sour stomach, and the ugly hypochondria!"[23] His wife Therese reported that his illness "colored the whites of his eyes"—they were yellow owing to elevated levels of bilirubin in his blood serum—"and his teeth [were] completely rotted."[24] In short, he felt "very tired," "as if his entire body had been beaten."[25]

None of this was an expression of sniveling self-pity. Forster aged more quickly than his friends. He saw his time as rapidly running out. He confessed to his betrothed that a "peculiar impatience" plagued him, that he was "already too old" for certain intense feelings—Forster was thirty. He was frequently haunted by premonitions of death, sometimes even "wished for death alone."[26] In his journal entry of September 22, 1784, this shocking note appears:

O I beg . . . that I may be no more and feel no more. Death! You must take me! I have done enough, in my twenty-first year I traveled round the world and am widely loved at thirty. It is time; it is time—Away with me! Otherwise, my glory will fade, perhaps my virtue, and certainly my pleasure, lost, before I even pass. I was never what one calls happy. A few times I dreamed, and—o how dreadful to awaken from that last beguiling dream! I cannot cheer my imagination again. The rapture of this air bubble, which so easily carries me up above all other tribulations in life and the world, has burst, and I am only just a worm trodden in the dust. The natural instinctual drive, which is so powerful in reconciling humans with life, appears not to have power over me anymore. . . . Out of madness I laughed today, leaping around the room and howling fragments of music from Re Teodoro; it burned and raged deep within me. Adieu![27]

When he became seriously ill in summer 1785 with "typhoid fever of the worst strain,"[28] he was relegated to a "complete dissolution of the blood"[29] and remained bedridden for seventeen days. Apart from a few cups of broth, he ingested nothing. "I cannot deny to you," he wrote to his friend Samuel Thomas Soemmerring, "that I would have died this time, with relish. God! It would have happened so quietly, so gently, so entirely without notice."[30]

All in all, it was a life broken into pieces.[31] "The voyage around the world was Georg Forster's life; the rest was a joyless appendix," Jan Philipp Reemtsma writes. But a life is not balanced. Confronted with the passage of time, Forster lived restlessly on, in spite of his malaise and occasional lachrymose moods. He had "no time to lose," he wrote in summer 1785.[32] How right he was. He had not even ten years left.

Nature's Balm

In spite of all its tribulations, nature proves to be a source of comfort for Forster. One look at the private notes in his journals reveals the power of natural impressions. Nature possesses, for Forster, a consistent attunement. It is not a neutral object of immediate experience but is intimately bound to the viewer's sensory perception. Forster loves "rich, bedecked nature." In an evergreen forest, he sees a "dark melancholy green"; he passes through a "laughing, but very flat area" and comes across a "dear sweet birch forest"; another time he endures a "nasty snow flurry." The sight of the Sudeten Mountains in the blue distance awakens in him a "dark, strange feeling,"[33] and the "whoosh of the wind in the uninhabited village of Hayn" is "lovely" to him, a "living breath . . . of nature." He says, "[It] reveals itself to us more closely, thereupon approaching our hearts more nearly, deeply."[34]

Nature's lessons promised relief: when Forster was upset and weary of everything, he embarked with Alexander von Humboldt on a journey along the Rhine. They "often spoke to the plants and the stones on the riverbank. . . . Their language is more instructive than the thick books that have been written about them."[35] When he was confronted with his melancholic temperament, nature alone bestowed comfort. Not cities, or museums, or, often, not even the people that Forster met on his journey brought him relief.

> Of all that I have seen up till now, nothing could stir me, could wake me from the gloom in which I was immersed, passing all day groaning in my plant life. But the lovely secret shade path, between the mossy stones, through the tall beech woods and poplars and fir trees, down along the trickling and rippling and babbling wild stream, ha! It closes in around me, it removes me from the tangle of the wide-open world before me, it presses me closely to the bosom of mother nature, alone and dark, but not horrid; rather, gentle, gliding steadily and silently, sweetly melancholic and dolorous, only the antithesis of the soul that delves into dark thoughts. In this way, nature knows to place balm and healing in the eternal variety of her works.[36]

When downtrodden, he enjoys passing through beech-wood forests, which appear to him "merry and charming" such that he could confess, "I feel myself strengthened and cheerful and joyful, and thus, babbling once again out of my feelings of gratitude."[37]

All of nature is there for Forster to perceive. He is rapturously bound up in it. He exposes himself to it with no inhibitions. Fortunate is he who knows it well: "O nature, what is more refreshing, and yet more permissible, than to love your works and to take joy in them!"[38] Tenderly he speaks of "dear mother nature" and confesses to his betrothed that "suffering has contributed much" to the development of his way of thinking. He "suffered through a great deal since youth," and this oppressed life has driven him "over to religious rapture."[39] Although he alludes to Rosicrucianism, which he felt bound to for some time,[40] ultimately it was nature revealing itself to him that served as a remedy, as salvation from his torments.

Physical Anthropology

In Christopher Martin Wieland's introduction in the *Teutschen Merkur* to Forster's *Voyage round the World*, he characterized the formula for Forster's felt closeness to nature and culture, to plants, animals, and people: "People are of course presumably products of nature, too." This assumption was outrageous for portraying a self-conception of human beings that was beginning to wean itself from the theological tradition. Should humanity no longer be understood primarily as God's creation? And if our existence cannot be metaphysically determined, then what are we? This insight into the naturalness of humankind led to the predicament of having to first establish a new conception of man. "Physical anthropology merits a particularly careful working out," Forster noted, when he was planning to write a handbook on natural history in July 1786. And three years later he proclaimed: "I so desire a new anthropology that it is practically for that reason alone that I am so eager to go to England. I hope to be the first to find *those* materials which seem vital to me for such work, namely *Nuances* of the human race, either living or in real portrait-like illustrations, as well as skulls that can be measured and compared." Before his travels with Alexander von Humboldt in 1790, he writes briefly of getting "hold of some findings" in the galleries of London and Holland to explore the natural history of "man and apes" and the "relationship of animals to people." Anthropology, which Forster seeks to provide a physical basis for, is no ancillary interest, no subfield of his research, but rather of central importance to his work. In the foreword to his collection of essays *Kleine Schriften: Ein Beytrag zur Völker- und Länderkunde, Naturgeschichte und Philosophie des Lebens*, the first volume of which was published in Leipzig in 1789, Forster explicitly emphasizes the "natural sciences in the broadest sense, and particularly anthropology," as being closely related to everything that he has written since his voyage around the world.[41]

The naturalness of humanity came into view in new and almost importune ways through the experiences that could be had on the edges of the known world. After all, what place were exotic savages, cannibals, or the innocent children of nature supposed to occupy in a history of salvation, which Christian Europe could still lay claim to but these strangers could not? As long as encounters with savages appeared to be exceptional—as with the discovery of America by Columbus—the indigenous people could be marginalized as nonbelievers and subjugated by missionary work. If, however, what emerged from the voyage around the world was that the majority of humankind did not fit within the parameters of Christian interpretation—what then? Humanity itself is called into question. "The proper study of mankind is man," Alexander Pope proclaimed in 1733.[42] This claim took on an ethnological timbre with the exploration of the world. The history of human salvation, as a situation endowed by God, began to be transformed into a cultural history, which presented itself in meandering variation—and insufficiently at that. In a form that had yet to be determined, human cultural history and natural history were to be joined together.

How reconciling nature with culture was supposed to work, however—in other words, the task of getting a physical anthropology of culture off the ground—was not yet resolved. And as always, when something new is still inadequate, one spares oneself from notoriety by putting a fresh coat of paint on the situation. With the publication of Jean Bodin's *Six Books of the Commonwealth* in 1567, Greek climate theory reemerged, and it was then revived yet again in the early eighteenth century.[43] In the fifth book, Bodin details the influence of geography and climate on social structures: "Political institutions must be adapted to environment, and human laws to natural laws."[44] Nature's influence on culture, it would seem, is determined by the aid of climate schemas.

To this end, the Greek climate model quickly proved to be too rigid. The distinction between cold, hot, and mild regions of the earth, which was common in antiquity, corresponded to the irreconcilable divide between the barbarians and the civilized. Only in temperate zones, according to Greek doctrine, can advanced civilizations develop. Ethnocentric climate theories like this, in which one's own nation is given the favored spot in the human hierarchy, were popular in the early modern period especially when it came to French authors. Yet, Jean-Jacques Rousseau was already calling into question, in his 1750 *Discourse on the Arts and Sciences*, the primacy of the scientifically civilized world over the *homme naturel*, or natural man. Rousseau was the first to account for the possibility that people from beyond Europe's narrow bounds of experience might be considered without the assumed declivity of civilization. Thus, Forster is able to describe "the nature of the people" on Tahiti[45] without resorting to Greek stereotypes. In spite of their exposure to a tropical climate, they have "a kind and generous temper," and their "character has pleased all the navigators who have visited them."[46]

The eighteenth-century voyages of discovery rendered the crude classification of civilized and barbarian obsolete. By the end of 1783, Forster was at work editing selections from Johann Peter Schotte's 1782 *Treatise on the Synochus Atrabiliosa* for publication in Lichtenberg's *Göttingen Magazine*. The final piece, which was published in 1784 under the title "Beobachtungen über das Klima von Senegal" (Observations on the climate of Senegal), contains a passage toward the end stating that Africans are "neither as senseless nor as irrational as commonly considered," even though the sun now and then causes "tremendous heat." This is a contradiction of classic climate theory, according to which independent reason could be cultivated only in temperate zones. Aristotle describes Asian peoples—as inhabitants of a hot zone—as natural underlings and slaves. By contrast, Forster can attest to the Tahitians' "innate freedom" and "natural grace."[47]

As attractive as climate theory initially seemed for determining the correlation between nature and culture, it did not accomplish what it professed to do. The Greek geographic worldview that it was derived from was straightforward by comparison but did not lend itself to being applied to the far reaches of the discovered world. Although Forster proceeds from the premise of a "*gradual* influence of the climate, which requires many generations," as its course is "slow but inevitable,"[48] he understood early on that the ethnologically observable distinctions between the world's cultures are the result of a variety of influences. "The different characters of nations seem therefore to depend upon a multitude of different causes, which have acted together during a series of many ages."[49] Bodin had already advised against being "confined solely to the consideration of climate," because "upon comparison of peoples within the same climate zone on the same geographic latitude, one finds four different skin colors, not to mention other traits."[50]

The basic idea behind Forster's *Voyage round the World* was to question cultural and political systems on the basis of natural conditions. Accordingly, Forster establishes a link between agriculture and societal systems in Tahiti and the eastern Polynesian Marquesas Islands. On the Marquesas, "the nature of their country, which requires a greater labour and culture than Taheitee, is one great cause of this difference; for since the means of subsistence are not so easily attained, the population and the general luxury, cannot be so considerable, and the people remain upon a level."[51]

Forster did not publish his handbook on natural history, which was to contain a physical anthropology, as planned. A preserved (but not datable) work plan for the handbook suggests a broadening of his perspective that forgoes simple ideas in favor of a variety of influential factors. "How much can be attributed to climate, local factors, special circumstances (e.g. constitution of state and religion)?" Forster asks. "Is it not extremely difficult to specify a national character; to say how nations are different from one another in what they can do?" In this schema, he abandons the latent determinism of popular climate theory without, however, contesting the obvious influence of natural conditions. Forster asserts what Herder had already argued, that humanity's freedom in relation to its natural origins is newly weighed and at once confronted by an obvious array of influences:

> Does it not often depend on just a few leaders for a state to maintain its constitution, which makes the greatest development of spiritual gifts and physical powers possible; and do the numerous linkages between foregone circumstances and events not have an effect on determining the direction in which these leaders play a part, at once imperceptibly and infallibly? How is it that not everything happening now is prepared from afar, such that we cannot think away the most insignificant link in the entire series without, at the same time, having to think of and expect a very different success from what in reality exists?[52]

In light of "the beautiful disorder to nature," philosophers inevitably fail in their attempts to "want to model nature after their logical distinctions." Nature has an edge, so to speak, which is effective in its diversity but not predictable. When Forster describes the Hawaiian Liwi bird, the red tree creeper, he observes that the length and curvature of its bill can vary considerably. The deviation from the supposed norm defies "all systems" and shows us "that nature does not always seem to work according to human concepts of order, harmony, and uniformity; consequently, our methods are only guidelines at best in her inestimable labyrinths." Nature's abundance exceeds our powers of imagination. Its reality is very different from "that which we bring to it, . . . our way of imagining it."[53]

If Forster did not write his handbook on natural history, it was likely for extrinsic reasons alone. Although his approach to understanding humanity in its naturalness shines through in his contributions to the controversies of his time, Forster would have seen that he had not enlarged the claim of an elaborate physical anthropology.

A Debate about the Human Race

The diversity of humanity in its cultural and physical forms can be counted among the modern gains in immediate experience. Ever since Columbus crossed the Atlantic, there seem to have been people in the plural, so to speak. Humanity, according to Forster, has become "everything nowhere, but something different everywhere." Humanity becomes visible in its amazing variety only where the imperial view has not eclipsed all that is foreign from the outset, and the question arises of which entity will prevail in this diversity. Forster viewed this question not as a problem for an arm-chair theorist but rather a matter of gained immediate experience. During his voyage around the world, he was astonished to find that the inhabitants of Mallicolo, an island in the New Hebrides, were "a race totally distinct," differing starkly from the other islands' inhabitants in customs and appearance. Forster suspected lineage from New Guinea and Papua. The differences were apparent, yet the explanation for their formation was difficult in the context of a biblically inspired worldview, which—decades before Darwin—still assumed a consistency of species and human creation by the hand of God. Forster admits in *Voyage round the World* "that our ideas were very premature on this subject, and that the history of the human species in the South Seas cannot yet be unraveled with any degree of precision."[54]

This new anthropological challenge must have caused every taxonomist anxiety, and Immanuel Kant was no exception in 1785: "The reports [*Kenntnisse*] that recent travelers are spreading about the manifold diversities [*Mannigfaltigkeiten*] within the human species have previously contributed more to stimulating the understanding to investigate this topic than to satisfy it."[55] The outstanding question was whether the unity of the human race guaranteed one uniform root for all humanity, given the variety of human forms, as evidenced by the diversity of skin color. How, then, is the plurality of human manifestations to be understood in the form of different human races?

The subjugation and atrocities committed in the name of racism render the concept of race unrecognizable in its original context of the European Enlightenment. The Enlightenment held the promise of "a rational scientific order of belonging in an increasingly complex world," "amidst an epoch like no other, which, from a contemporary standpoint, places traditional belonging at our disposal."[56] Carl von Linné had already introduced for the plant and animal kingdoms binomial nomenclature, according to which every classification was composed of two names, identifying the genus and species. The concept of race was supposed to accomplish something comparable by differentiating the unity of humankind and helping to show what the human race in its diversity has in common.

What might look like a problem within the scientific community itself contains tinder for a larger controversy. In 1785 Kant published the essay "The Determination of the Concept of a Human Race" in the *Berlinische Monatsschrift*, primarily to disambiguate the term. Forster's reaction to Kant not only sparked a dispute but verged on polemic in its attempt to lay bare the political implications of the debate. Meticulous care was taken to bring taxonomic clarity to this argument between Kant and Forster.[57] Nevertheless, the controversy between the two remains curious. Forster misunderstands Kant's position at points, while Kant permits himself judgments about people on the other side of the world. Thus it is unsurprising, according to Ludwig Uhlig, "that the debate on both sides led to fanciful hypotheses and ended in Aporia." The vehemence of their argument might today be regarded with dismay, as Manfred Kühn concludes: "Large parts of this discussion must strike today's reader as boring at best and offensive at worst."[58] The debate between Forster and Kant took place in the crosshairs of the era's clash between experience and systems, empiricism and speculation, ideology and science. Determining the origins of humankind was no trivial matter.

What were their positions? A full decade before his essay "The Determination of the Concept of a Human Race," Kant presented his first attempt at systematically measuring and reasoning through the unity and diversity of the human race. In "Of the Different Human Races," published in 1775, he takes the position that "all human beings belong to a single lineal stem stock . . . in spite of their differences." Nevertheless, there are four races: "1) the race of *whites*; 2) the *Negro race*; 3) the *Hunnish* race (Mongolish or

Kalmuckish); and 4) the Hinduish, or *Hindustanisch*, race." The formation of these four races is assumed by Kant to have germs that qualify them to adapt to the most different environmental conditions: "Human beings were destined [*bestimmt*] for [living in] every climate and any condition of the land. Consequently, various germs and natural endowments must have laid ready in them to be at times either developed or held back so that they might become fitted in a particular place in the world and seem, as it were, in the succession of generations, to be native to and made for [these places]." The differentiation into four races is thus precipitated above all by climatic influences but follows the predispositions that already exist naturally in human lineage. In this way, Kant is able to claim the diversity of races and the unity of the human race at the same time. Thus, although "*Negroes* and *whites* are certainly not different kinds of human beings (since they presumably belong to one line of descent), they are two different races."[59]

In his subsequent essay "Determination of the Concept of a Human Race," which sparked a reaction from Forster, Kant undertakes "to define [*bestimmen*] precisely this concept of *race*,"[60] as it comprises "first, the concept of a common line of descent, [and] second, [the] *necessarily heritable* characters of the class distinction of the descendants of the [line] from one another." By "class" distinctions, Kant means the four human races that are enabled by embryonic predispositions. The characters of a race are thus, necessarily—if developed at all—of a consistency that is fundamentally guaranteed by heredity. Again, Kant defends the unity of the human race: "The class of whites is not distinct from blacks as a separate kind in the human species. There are absolutely no *different kinds of human beings*." Kant persists: There is *one* human, who is able to adapt to the most different conditions, as he expressed in a handwritten note: "Can live in all climates and all kingdoms of nature."[61]

Kant later spoke of this second essay as "a little essay."[62] Insofar as he could classify it, the theme of the human race was not unimportant to Kant, but it was not of central significance for him, either. Ultimately, his universal moral philosophy, based on the asset of practical reason, remained completely unaffected by the differentiation of races.

Forster accepts none of Kant's arguments. In a letter to his friend Soemmerring, Forster writes of Kant's essays and the theory of race represented in them, "against them I have infinitely much to object." First, Forster defends the essential value of his own immediate experience against the claims of an "armchair philosopher." Kant must have struck him as a swaggering philosopher, going on about things for which he has no corresponding experiences at his disposal. Kant claimed, for example, "We still cannot form a reliable notion of the primary color of the South Sea Islander on the basis of all the descriptions [we have been provided] up to now." In his response to Kant, which appeared in the *Teutschen Merkur* in 1786 under the title "Something More about the Human Races," Forster writes—arguing a bit off the cuff—that the answer can be found "recounted definitively and to the same effect in the more recent travel descriptions," namely his own. Pages of exposition follow about the observable nuances in skin color among the most different peoples. Forster thus draws a clear methodological demarcation, which is reinforced by the accusation—expressed in Forster's letters—that Kant's theory of race proceeds "from his sphere," in other words, "from his field" as a philosopher, and thus, Kant "involves himself in things" of which he has no understanding.[63]

Kant, by contrast, insists on having necessarily "determined well beforehand the concept we wish to illuminate through observation before we for its sake turn to experience," because it is certain that "nothing purposive would ever be found [in nature] by means of purely empirical groping about without a guiding principle that might direct one's search." Forster defends the immediacy of his experience, which requires no prior conceptual disciplining. "To the extent, therefore, [that] the impartial observer only faithfully and reliably reports what he perceives without pondering for a long time which theory [*Spekulation*] his perception favors, I would look for instruction more confidently from him than from an observer who [has been] tempted by a faulty principle that lends the color of his glasses to the objects [he is investigating]."[64] Forster knows about the blindness of those who "raised a general cry after a simple collection of facts" and with those facts "received a confused heap of disjointed limbs, which no art could reunite into a whole."[65] But nature is so staggeringly diverse for Forster, the world so abundant, that "everything is connected through the finest modulations";[66] yet, "fallacious systems" threaten to force the experience of nature into a corset.[67] The "rubbish" of logical distinctions is indeed "truly more harmful than useful."[68] Forster braces himself with all his might against the methodological constraints of a rationality that flouts observable phenomena. All terms, systems, and classifications are to follow from one's perception of nature, not guide it. Forster means to be able to make up for Kant's arrogance, which seeks to impose his system of thought onto the diversity of nature. "In a word, the order of nature does not follow our divisions, and as soon [as] we want to force these [divisions] on [nature], we lapse into absurdities."[69]

The formation of different human races can be observed by long, ongoing, climatic conditions no more than experience can trace the human race's origins to one lineage. Kant draws a strict distinction between descriptions of nature and natural history, whereby the latter is a "separate science" that has yet to be substantiated—a distinction that Forster is only conditionally prepared to accept "if only both are over and over again united and treated as parts of a whole."[70]

With every fiber of his being, Forster defends to Kant the advantage of immediate experience, knowledge derived from experience or the experiences of a lifetime, and feeling over reflection, over a classification system that creates order, over abstract terminology, in short, over the "spawn of reason."[71] Kant wants to understand how reason constitutes the reality with which it is concerned; Forster wants to see the immediate impression defended. Kant is a transcendental philosopher, because he uses experience to examine the condition of possibility; Forster would probably describe himself as a realist. "For me reason is a living mirror of reality," Forster writes, unencumbered by the philosophical *niveau* of his time. Reason for him is a faculty that is receptive, not constructive, because it "reflects back the radiating out of what is real."[72] Reality should be purely reflected in people. Thus, Forster stands up for the primacy of his impression of experience and claims to subordinate all "concepts, ideas, or representations." "As concerns me, I have no conviction more vigorous than this," he writes, "that no representation models what is real as being real to me, because the immediate effect of what is real cannot be denied to me, and precisely because of this, I *feel* the presence of this realness so undeniably where it is, outside of myself."[73] Compared to Kant's critical reflections, which might seem philosophically naive, Forster was aware of the unassailable advantage of his experience, which he was not prepared to sacrifice for the consideration of philosophical objections.

For this reason, he does not shy away from calling into question Kant's approach to monogenesis, human origination from one lineage. Forster considers instead the possibility of a plurality of local origins, whereby human ancestry can be traced back not to a single lineage, but to at least two: black and white. The difference between them is not great, as he emphasizes: "The Negro who is most like the ape is so closely related to white human beings that the distinctive properties of each interweave and fuse in the interbreeding of these two lines of descent with one another in the half-breed. The divergence is very small. The two human [forms], the black and the white, stand very closely next to one another."[74] Why speculate about polygenesis, though?

The hypothesis that humans developed from different lineages certainly had more proponents. In *Sketches of the History of Man*, published in 1774, Henry Home considered the view that there could have been a "local creation" of man "by the hand of God," in other words, human creation in different forms in different places. Soemmerring, too, approaches the hypothesis as an anatomist when he writes, "Africans and Europeans differ not in variety but in species, and there were—as I like to say—two Adams."[75] Soemmerring's point gets at what makes this discussion so controversial: polygenesis stands in opposition to the biblical creation story.

This must have riled Forster. Kant, after his "Determination of the Concept of a Human Race," wrote a historical-philosophical study, titled "Conjectures on the Beginning of Human History." In analogy to a natural history, he develops the genesis of a morally rational world, in which humans develop from nature and perfect themselves through reason. As a point of reference, Kant chooses the first book of Moses, chapters two through six; yet he does not want to depend on theological assumptions in his use of reason. Forster mistakes Kant's use of tradition as a narrative framework for his position. It bothers him that Kant "resorts to curious biblical explanations, clearly seeking out a new point of view for the Mosaic scriptures, which everyone knows and honest religious scholars wish to bury into oblivion." Forster can hardly believe his eyes: "To speak of the Moses-Kant metaphysics is indeed the worst thing I can think of."[76]

In their debate about race, Forster seems to want to represent modern science, by implying that Kant is still dependent on metaphysical and theological assumptions. He sees it as an explicitly aimed "attack on this one old book," the Bible, "if one presents a possibility of more than one human line of descent." Yet only Forster weighs a "wise dispensation [*Fügung*] of providence" against any judgment of humankind around the globe,[77] while Kant—quite distinctly in his *Critique of Judgment*—shows teleological assumptions of this kind to be one of the properties inscribed into our capacity for cognition.[78] In no way does teleology prove to Kant that "there is such an intelligent Being," who has determined, in one act of providence, the course of the world. His treatment of theology proves nothing other than that "according to the constitution of our cognitive faculties and in the consequent combination of experience with the highest principles of Reason, we can form absolutely no concept of the possibility of such a world as this save by thinking a *designedly-working* supreme cause thereof."[79] We can only suppose a teleology, but this is not to say that reality is actually determined through providence.

Yet, Forster had seen for himself that Kant was not so easy to argue with. So he cannot accuse Kant's claim of monogenesis as being a metaphysically—or worse, theologically—inspired hypothesis that lacks experiential basis, without, in the same breath, advocating for polygenesis. But how to empirically prove the multilineal origins of the human race? Strictly speaking, Forster does not do so, even if he is, at times, rashly viewed as a fierce advocate for humanity's multiple origins.[80] In his response to Kant, he offers polygenesis merely as an equitable possibility without answering "decisively in the affirmative." It is no less "improbable nor inconceivable" than the assumption of monogenesis—"however, everything is admittedly uncertain."[81]

Why the debate, then? In an undated note, presumably from 1786, titled "Human Races," Forster explains his concerns with regard to Kant:

> I only wanted to show that the matter can be considered from another point of view, too, and that it can be spoken about without apodictic certainty. If the matter is to be sorted out, its pros and cons will have to be aired. That humankind is one species will be established as soon as the concept of a species is determined in such a way that lets it be established. Whether we all descend from a single line, however, is nowhere near a consequence; quite the opposite, as far as I am now in a position to judge, it will always remain a matter of faith, i.e., a matter about which one can, at most, only speculate and find preferable to others.[82]

For his critique of Kant, Forster chooses the stylistically open form of an essay written as a letter, rather than a formal paper, and not without reason.[83] In a letter to Herder dated January 21, 1787, Forster puts his cards on the table and emphasizes that he is "in all seriousness still a long way from believing that the human race truly has multiple progenitors. I only think the matter has made gains in being seen from another side and that this shows how little proof there has been up till now towards accomplishing what it claims to do."[84]

One thing cannot be overlooked: Forster sought out the debate with Kant. He "dared Kant"[85] to refute his definition of the human race "in all possible ways, without wanting to provide another classification of the human race himself." For Forster, it was not about "putting forward a new opinion, but rather refuting the incorrect Kantian one; the rest is incidental and window dressing."[86] Forster had no fear of bias. In a March 1790 letter to Sophie von LaRoche, he describes himself as an author who "has not yet learned to enthuse over his skill" and who, "in the manner of young writers envisages his subject matter very clearly from *one* side," even if there are several sides to consider.[87] His debate with Kant is underpinned by this intended emphasis on one-sidedness. Their public dispute required a rivalry of positions that necessarily diverge in order for the pros and cons to emerge.

As a writer, Forster never lost sight of his audience. He wanted to be publicly recognized. His argument with Kant—as his letters verify—was supposed to garner popularity for him while he was stuck in Polish exile. "I confess," he writes to Soemmerring, "that I am curious what the public will say; because no one has heard me speak in this tone before." And in the event that "the public should confer upon me that I am right, right against *Kant*, then, that can be chalked up as a win, too."[88] His publisher Spener specifically pointed Forster to the Kant essay in *Merkur* and urged him "to be attentive to the impression that he makes with people whose judgment has weight in the academic world; and to furnish me with information about it."[89]

For Forster, though, this was not at all what it was about. After all, for him the provocation was that Kant "said many inaccurate things particularly about the South Sea islanders."[90] However, if it meant gaining public attention, by way of a dispute with *the* German philosopher of his time, nearly any means suited him—or to put it another way, he did not once let himself be deterred by the fact that his knowledge of Kant's philosophy was only "through the third man," insufficient and merely secondhand. Thus, Forster named the debate with the Königsberg philosopher the "Klopf-fechterstreich"—in fencing schools it was normal for students to show what they were capable of in an exhibition match, in which, according to the dictionary compiled by the Brothers Grimm, it came down to "more noise and crossing of swords than art."[91]

In retrospect, Forster himself saw that his dispute with Kant was argumentatively flawed. For Kant's part, in the article "On the Use of Teleological Principles in Philosophy," he explicitly responds to Forster and urges clarification of the epistemological conditions at stake in the question being debated, which clarification threatens to be embarrassing for Forster. "From a lack of prior philosophical knowledge, and almost even more, because I do not understand philosophical jargon," Forster writes Jacobi, "I fell into an argument with Kant and now run the risk of publicly being thrown from my horse, because he uses his artificial language to curl up into the most invincible, prickliest form of hounded hedgehog to make you believe you cannot get at him." Forster seems weary of the debate, finding "Kant's verbiage tiresome."[92]

Overall, Forster comes off as having instigated the debate with Kant. He sought it out because—cut off in Vilna from the academic world—he needed a publicized success. He engaged in a debate about the human race even though he did not "love the word *race*," because it already connoted too much of a preconceived structure as to the diversity of peoples; he used it "in order to use Kant's word." Forster introduced polygenesis not least because Kant claimed the opposite. Moreover, polygenesis brought him into alliance with Herder, who, since Kant's vociferous critical review of Herder's "This Too a Philosophy of History of the Formation of Humanity," had his own reservations about the philosopher from Königsberg.[93]

Despite this argumentative heteronomy, the debate is of great substantive weight for Forster. First, it comes out of his attempt at physical anthropology, which he never did get a handle on, but he did not completely give up his work on it, either. Second, the question of the unity and multiplicity of the human race possesses a moral and political grounding that surpasses the scientific question regarding the determination of the human races. The impetus for this may have originated in part from Soemmerring, who, in his 1785 book *On the Corporeal Difference between the Negro and the European*, sought to identify anatomically verifiable differences between human lines of descent. Forster was in the audience for his friend's anatomy lecture in Kassel,[94] in which Soemmerring considered whether "the blackamoor" was allotted "a lower step to the throne of humanity." Of course, Kant's statements were not free of disparaging remarks, either: He spoke of "the strong smell of the Negroes, which cannot be avoided by means of any [degree of] cleanliness,"[95] such that "all Negroes stink." In his classification of the human races, he described the "Negro" as "lazy, soft, and dallying,"[96] while granting whites the privilege of remaining the closest descendants of the original human lineage.[97]

In light of such assessments, Forster's argumentation possesses a deeper political layer. Toward the end of his essay "Something More about the Human Race," Forster debates, without even remotely addressing his dispute with Kant, whether the hypothesis of polygenesis or monogenesis is in a better position to prevent or advance the disgrace of human enslavement. This "cursed bondage," as he writes Therese, is an "ugly thing," because it "poisons the roots and blunts human feeling."[98] Surprisingly, Forster doubts that the thought of monogenesis "ever, anywhere, even once, caused the raised whip of the slave driver to be lowered." On the contrary, "the idea of a second human species presents itself to the highest understanding as a powerful means of developing thoughts and feelings worthy of the understanding of a *rational* earthly being."[99] Why should it be this way? Forster is not able to fully flesh it out. Because if one assumes different lines of descent and—as Soemmerring did—treats it as proof that "generally, on average, the Negro veers somewhat more closely to the primate family than that of the Europeans,"[100] then, a human scale of worth is suggested, which in turn would seem to suggest a different ordering of the races. Although the possibility of enslavement is not entertained, racial discrimination is not ruled out, either.

Forster himself does not seem to entirely trust his idea that the hypothesis of a second human species might give way to political humanism. Because this argumentation is peculiar, to put it mildly, and prone to misunderstanding,[101] Forster himself was dissatisfied in retrospect with his contribution to the controversy. When in 1788 he published a volume of his most important essays under the title *Kleine Schriften: Ein Beytrag zur Völker- und Länderkunde, Naturgeschichte und Philosophie des Lebens*, he did not include his response to Kant. And when, that same year, he wrote his essay "Über Leckereyen" (On delicacies), he only mentioned the problem of human lineage and avoided developing it any further.[102] He retreated from the debate.

Only in his zoological lecture, "Praelectiones zoologicae," which he delivered in Vilna in 1787, did Forster explicitly return to the question of polygenesis. This lecture played no part in his dispute with Kant, since the Königsberg philosopher could not have known what ideas Forster presented there. It is interesting, nevertheless, as it documents how quickly Forster abandoned the position of multilinealism. The lecture, and the compendium underlying it, refers in large part to the sources consulted, which are occasionally reported verbatim, as was usual for that time: in this case, Carl von Linné's *Systema naturae* and Johann Friedrich Blumenbach's *On the Natural Variety of Mankind*. Only occasionally does Forster deviate from these templates. First, he determines that there are *Exempla* of human beings—he counts seven in all—in other words, permanently different human races.[103] Moreover, there are *Varieties*, which is to say, temporary variables to the constant *Exempla*.[104] Departing from his discussion with Kant, Forster now emphasizes that, when forming judgments on the original varieties, all we have are suppositions; if one does not want to rely on hypotheses, we must enumerate the *Exempla* that are known today insofar as they are apparent to us, and in doing so, we name the differences evident in their forms, as well as their singular and characteristic marks.[105] In his lecture, Forster manages to turn a hypothesis about human development into a descriptive morphology.[106] We may be able to document the differences of humankind, but we know nothing about its origins.

It is apparently not difficult for Forster to backpedal on his dispute with Kant. After all, Soemmerring held firm to the premise that "Negros" are "for highly probable reasons—which natural history, physiology, philosophy, and written reports present to us—sprouted from a common patrimony with all other people." Rooted in experience, Forster's anthropology resolutely and repeatedly stresses the unity of different peoples and groups of people, despite all differences and the transparent scale of civilization's strata.[107] "The same instincts are active in the slave and the prince," he writes in *A Voyage round the World*.[108] For Forster, there is only *one* humanity and not different kinds of humans, even if he factors multiple lineages into the genesis of humankind.

Were Forster and Kant ultimately that divided? Was the gulf between them so insuperable? Forster later sought rapprochement with his opponent. He declared his veneration for the "excellent Kant" and communicated to one of the philosopher's students that his own essay from that time had a "touch of polemicizing spleen, which I wished to take back soon after I saw it published because it neither pertained to the matter at hand nor was it befitting toward a man like Kant." All of his publications from this period in Vilna were of this ilk, a consequence of "physical ailment," which may indeed be correct.[109] Forster summed up his time in Vilna as a "period of melancholy and semi-madness," and on another occasion he apologized for his remarks having been "a bit sharp," a tangible "consequence of expelled fluids,"[110] in other words, the result of his fever. When Kant's *Physical Geography* was published in 1802, a laudatory reference could be gleaned to Forster's "instructive remarks" about "matters of the physical description of the earth."[111]

In spite of this conciliatory note, Forster's opposition to Kant's espousal of monogenesis can be seen as an attempt at safeguarding human diversity. Forster accuses monologic reason of imposing its model of cognition on phenomena.[112] In his correspondence with Christian Gottlob Heyne, Forster even speaks of a "despotism of thought" that he finds characteristic of Kant's philosophy. There is no absolute truth; rather, "everything true that a person can put forward" is only "relatively true."[113]

In spite of any questionable arguments, Forster's criticism of Kant is guided by a shrewd intuition, as Kant did in fact pursue the idea of a progressive process of civilization, which was Eurocentrically conceived. According to Kant, Europe would make "laws one day for all others probably." An ideology of development such as this threatens to ensnare "primitive" peoples in the victim's estimation. In his review of Herder's *Ideas for a Philosophy of the History of Mankind*, Kant questions whether the author might be suggesting "that if the happy inhabitants of Tahiti, never visited by more civilized nations, were destined to live in their quiet indolence for thousands of centuries, one could give a satisfactory answer to the question why they bothered to exist at all, and whether it would not have been just as well that this island should have been occupied by happy sheep and cattle as by happy men engaged in mere pleasure."[114] The sheer moral perfection, which must commence with Europeans, justifies the existence of exotic peoples in perpetuity! Forster grasped that the danger of enlightened reason lay in its refusal to be challenged by experience. Although his debate with Kant was by all means unsuccessful, there are flashes of Forster's valuable core idea of political humanism throughout: humanity necessitates upholding diversity.

In "Darstellung der Revolution in Mainz" (Account of the revolution in Mainz) written in 1793, Forster returns to the topic of his dispute with Kant one last time. He does not know, he prefaces his musings, "whether the same virtues and the same weaknesses develop without any variation in all places on earth" and whether "different places can produce exactly the same people."[115] Forster does not mention his theory of polygenesis again, but he insists on the connection between humanism and diversity. Why is it, he asks, that "people from all places are poured as if into one mold?" His personal predisposition may be freely deployed, but in no case should a restrictively "precise conformity by the citizens of a place" be called for—this can be read as a late echo of his controversy with Kant. The philosopher had indeed claimed in his review of Herder's *Ideas* that the work was lacking "a logical precision in the definition of concepts."[116] Forster had every reason to turn Kant's rebuke of Herder on himself, as their debate reveals. Still, Forster's later defense of human "imprecision" reveals just how uneasily Kant's classifications sat with him.

From our perspective today, Forster's dispute with Kant anticipates postmodern debates about the Eurocentric ideals of enlightened reason. The experience of diversity should be defended against the monologue of reason. With all his might, Forster points out the "beautiful phenomenon of diversity in the human race," which he understands as being endangered.[117] Sarcastically, he notes "that many hypotheses would have better standing if the ugly blacks could be demonstratively removed entirely from the South Sea." It depends on whether we accept the person who cannot be integrated into our thought patterns. "They are, however, now once and for all there," Forster writes, a laconic remark about blacks and their integrity in the face of European illusions of unity.[118] He sees their integrity as being worth protecting—even if his arguments are dubious.

Political Sheet Lightning: Cook, the Statesman

James Cook was expected to return to England during the course of 1779. He had set out three years earlier on his third circumnavigation of the earth with the purpose of finding a northwest passage between the Atlantic and the Pacific. "Cook has not returned, nor has anything been heard of him," Forster writes to his publisher Spener in September 1779.[119] It was not until the January 10, 1780, issue of Anton Friedrich Büsching's newspaper *Wöchentliche Nachrichten von neuen Landcharten, geographischen, statistischen und historischen Büchern und Sachen* that Forster learned of the fate of his erstwhile captain: James Cook had been murdered by native Hawaiians on February 14, 1779.

His death came as a shock. Although Cook had once taken seriously ill during the second circumnavigation, his death was unfathomable. George Gilbert's travel journal bears witness to the traumatic effect that Cook's murder had on those who remained on board: "On the return of the boats informing us of the Captains death, a general silence ensued throughout the ship for the space of near half an hour: it appearing to us somewhat like a Dream that we could not reconcile ourselves to for some time." Now the European readership demanded information about the circumstances of Cook's death, and the task fell to Forster to recount the events for the German reading public. Forster's "Fragmente über Capitain Cooks letzte

Reise und sein Ende" (Fragments on Captain Cook's final voyage and his end) was published in January and February 1781 in Lichtenberg's *Göttingisches Magazin der Wissenschaften und Litteratur*.[120] He also translated the account of Cook's last voyage, which was released in two volumes in 1787 and 1788: *Captain Jacob Cook's Third Voyage Round the World, undertaken and performed by royal authority in His Majesty's Ships the Resolution and Discovery to the Pacific Ocean and the North Pole, 1776–1780*.

As early as December 1786, Forster mentions the "agreed-upon introduction" by his publisher Spener to the book about Cook's final voyage of discovery—referring to Forster's *Cook, the Discoverer*. By April 1, 1787, he announces its completion.[121] With this essay, Forster is provided the opportunity, and the temporal distance, to appreciate Cook's person and accomplishments. This text is so important to Forster that he includes it in the first volume of his *Kleinen Schriften* in 1789 as the first piece in the book.

The essay—which was decidedly not called "Cook, the Conqueror," so as to distinguish it from Columbus and to appreciate Cook's enlightened acts of discovery—is remarkable. Even Forster understood it to be a special text. "Despite the all-consuming effort, with which I have worked on this essay, I have never done anything with it; my wife believes it will cost me my life," Forster writes in January 1789. At first glance, the text offers a "Panegyric"[122]—a "eulogy" and "memorial"[123] to Cook—a necessary tribute to the leading explorer of his time for his pioneering deeds and character traits. Appropriately, the first printing bore the title *Cook, the Discoverer: An Attempted Memorial*. Forster did away with the subtitle on the subsequent printing of his *Kleinen Schriften*.

For Forster, remembering Cook also meant reflecting on his own youth: "Here I cast a glance at the distance covered, and even I marvel at a voyage in which I took part, but which, thirteen years later, appears like the wondrous events of a dream." Forster speaks of the "paternal care" that Cook afforded his underlings—in retrospect, he may have elevated Cook into a surrogate father and conceived of his essay as a son's tribute of gratitude. Jan Philipp Reemtsma has spoken of an "adoption in spirit," in which the agony of writing indicates an underlying psychological conflict.[124]

Yet, the text surpasses its occasion of honoring Cook; Forster uses the essay as an opportunity to "unpack [his] philosophy." Beneath the surface, *Cook, the Discoverer* opens a window onto Forster's philosophy of history, offering an inkling of the trials to come and a pronouncement of revolution. Forster wrote a political doctrine of revolution into his essay *Cook, the Dis-*

coverer, encapsulating the circumnavigation of the world with a view to the looming contemporary horizon. "The scope and range of a great voyage of discovery together with the variety of things which occur, results in the account being read with profit from as many points of view as there are modifications or branches of human knowledge."[125] This is to say, the essay can also be read as an object lesson in political philosophy. The text recovers moments of a germinating revolutionary consciousness, which makes it a fascinating document of the period just prior to the French Revolution.

First, the essay emphasizes Cook's accomplishments that the reader may be expecting would be paid tribute to. Before Cook, geographical knowledge of the world was in many respects a "chaos of unsubstantiated opinions." After three circumnavigations, "now the globe is known from one end to the other. Can anyone who looks at a map and sees the changes in geography, achieved through the passionate research of a single man, doubt for a moment that our century could compete with the greatness of any age?" More decisive for Forster, though, is "the value of these discoveries."[126] He ranks Cook among the heroes of civilization's history of progress, which requires "subjective perfection" of humankind for the purpose of "a more complete knowledge of truth."[127] What Cook gained in knowledge from his voyages was not only of geographic, ethnological, and natural-historical interest; rather, his discoveries were going to have "deep roots and long have a decisive influence on the activities of mankind."[128] Cook is turned into a scientist for pragmatic purposes. If one still has in mind Forster's controversy with Kant over the methodological importance of immediate experience, it becomes apparent that *Cook, the Discoverer* was written tacitly against Kant.[129]

As Jörn Garber writes, Forster turns Cook into a "protagonist of scientific discovery," providing him with "all the attributes of a promethean founder of civilization and science" and linking this image to "an energetic rejection of the Rousseauism of his time."[130] "Rousseauism of his time" refers to the idea of a meek and noble savage, unspoiled by civilization, who, in his naturalness, is far superior to the European who is defiled by the excesses of societal deformation. For Rousseau himself, this was all conceived as a thought experiment, beyond any factual claims, for the purpose of critiquing civilization. But the prospect of finding a "primitive state" of humanity might have secretly played a part in the world travelers' expectations. Forster does not succumb to this perceptual template, even though he did indeed encounter people in a "natural" state, as Europeans were wont to imagine. On Malekula, an island east of Australia (also called Mallicollo),

Forster came face to face with its indigenous inhabitants, who, on the basis of their foreignness, "provoked us to make an ill-natured comparison between them and monkies." But the operative word here is "ill-natured," and Forster seizes this opportunity to rail against Rousseau, who had so neglected the intelligence of these people "as to degrade themselves to the rank of baboons."[131] Forster draws a strict line between primates and people but maintains a blurring of distinctions for the "ill-natured," or "boshaft" as he put it in the German edition of *Reise um die Welt*. To him, there are no pure, natural people, or, as he concludes in *Cook, the Discoverer*, "there is no mere animal stage of nature."[132] The traveler "who roams all four corners of the globe, fails to find any of the lovable people he had been promised in every glade and wilderness." Forster is critical of the idea of "the foolish child of nature" and finds the Rousseauian concept of an unspoiled primitive state to be nothing but "prattling fabrications."[133] The savages, he continues, are today our friends, tomorrow our foes; they are not to be treated "according to the ideas in which one has been schooled," as they do not conform to our clichés.[134]

Although Forster's desire to separate himself from Rousseau and all the "superficial philosophers who re-echo his maxims" may misunderstand the philosopher's speculative approach,[135] that desire still serves to expressly defend the possibility of a progressive history of civilization and to present Cook as the patron of such progress. For Rousseau, the history of modern society is a history of decline, and the recourse to an imaginary primitive state serves as a foil for conducting an anamnesis to the pathology of contemporary culture. In this way—as in psychotherapy, where a patient's trauma is revisited so as to introduce an alternative that frees the development of the personality from its pathologies—it is thought that a society, freed from its undesirable developments, can become possible. From these subtle, not unsophisticated, considerations grew the widely held notion that Rousseau accepted as fact not only the existence of a primitive state, but that such a state can be returned to. "Back to Nature!" became a motto, though it is not found in any of Rousseau's writings. Nevertheless, it is precisely this alleged decree from which Forster seeks to separate himself. The most remarkable formula of his opposition to this brand of Rousseauism can be found in his short essay "New Holland and the British Colony in Botany Bay." The critique of bourgeois society, he states, fabricates "a contradiction between nature and culture." Forster, by contrast, views culture as emerging from nature, as possessing an inseparable, natural foundation. The

Scottish philosopher Adam Ferguson, who perhaps influenced Forster, had already grasped the naturalness of human art, which is to say, its overall cultural technology, in his 1767 "Essay on the History of Civil Society": "We speak of art as distinguished from nature, but art itself is natural to man."[136] For Forster, "cultural advancement" is now "in the interests of mankind." Thus, it is to the credit of "the immortal *Cook*" that the advancement of geographic knowledge from pole to pole could be leveraged into the progress of human culture, too.[137]

In contrast to the armchair explorers, "where the golden pen does not founder on any reef and the paper ocean is not disturbed by any waves,"[138] Cook's worldview was steeped in experience.[139] The basis for his judgment was "free of the fetters of prejudice," and he "had not studied the savages theoretically at all." Forster undeniably fulfilled the requirement of writing an obituary that emphasizes positive attributes; he stated, for example, that Cook "punished rarely and unwillingly, never without a pressing need and always in moderation."[140] The character portrait that he paints of Cook, however, possesses a political foundation that exceeds the parameters of an obituary.

According to Forster, the principle that guided Cook's actions was "his respect for human rights,"[141] which—in his authority to make hierarchically oriented decisions on board—honored "humanity even in a common sailor." Compared with some "naval despots" who "consider their arbitrary will as the highest law," Cook was always attentive to signs of appreciation for the sailors: "The duty officer stayed on deck, wet and rigid with cold, for the full duration of this watch, and Cook ate the same food as the common sailors."[142]

Cook's voyages are covertly turned into an allegory for the true task of the state. Only at first glance is it surprising to stumble upon a sentence in *Cook, the Discoverer* that steers the entire text into an enlarged semantic field: "To protect Man and to render him happy are the two great problems of statecraft."[143] A few years later, in 1793, Forster wrote an essay with the title "Über die Beziehung der Staatskunst auf das Glück der Menschheit" (On the relationship of statecraft to human happiness), in which he urged princes and priests to "clear away the barriers that hinder the free development of our powers," for which human dignity is the "proscribed guide of life."[144] Yet, already in his obituary for Cook, Forster hints at a "change in conditions."[145] "The progress of enlightenment," fulfilled by Cook, "gives a strong momentum to that activity which is the chief condition of the occupation of mankind because it brings about new conditions."[146] That is to say, Cook's exploration of the world is not unimportant for the development of the political. On the contrary, among the remote islands of the South Seas, the voyage revealed "that humanity in the hot countries should be the first to succeed in attaining such a beautiful bond of companionship." Forster, of course, views the equality of the human race as having developed under the sun of the South Seas. Where this bond of companionship was disturbed by European subjugation and exploitation—in his essay "Über Leckereyen" Forster mentions the "Negro trade, in order to be able to enjoy a few delicacies, like sugar and coffee"[147]—it emerges that the colonies "as soon as they are able to exist by themselves . . . become emancipated and sever their connecting with the parent tree." Unsurprisingly, Forster is cautious about dating anticipated events when he writes, "Within a few centuries an important apparition will become visible in the political sky in the west," even in Europe.[148]

It likely would have come as a surprise to Cook to find himself cast as a statesman by Forster. Extolled by Forster as the vanquisher of scurvy, Cook is turned into a "saviour and deliverer from this cruel and wasting death" and depicted as "the protector of the lives of many thousands who would henceforth be able to sail the oceans in health and confidence."[149] He thus fulfills the task of delivering happiness to humanity through statecraft. For this virtuoso feat of rhetoric to be successful in presenting Cook as a statesman, the ship at his command must be metaphorically turned into a ship of the state. Only then can his key doctrine on political society be read in the circumnavigation of the globe. When the *Resolution* became stuck on a reef off the shore of Tahiti, "at this moment of common danger everyone, regardless of rank or whatever task he might otherwise have been engaged [in], lent a hand in winching the ship off the rock and into deeper water. Surgeons, astronomers, naturalists and draughtsmen, all of them people who normally have nothing to do with running the ship, were panting away at the capstan in temperatures of more than thirty degrees." The camaraderie on board the ship is turned into a symbol for solidarity, leveling the men's differences in rank. "Experience incontrovertibly teaches us that humanity either succumbs under an intolerable burden or shakes it off with just indignation."[150]

To play the role in the civilizing process that Forster intended for him, Cook needed to not come from privileged circumstances. He was born the son of a day laborer on October 27, 1728, in Marton, a village in Yorkshire, where he worked as an apprentice at a general store before going to sea. He found employment with a shipping company that transported coal from the north of England to London—not easy work on the rough waters of the North Sea. Forster emphasizes that Cook served "as common sailor and mate of a merchantman in this tough service, enduring the sort of hardships and adversities which he later tried to palliate for his subordinates."[151] He is stylized by Forster into a working-class hero who works his way to the top through his toils. Cook goes on to become a captain in the Royal Navy, where his humanity is nourished by the experiences he shares with the subordinates under his command. Any improvement in their conditions—Forster's biographical object lesson concludes—was mobilized from below, not by the uppermost reaches of society.

No more can be expected of the nobility, certainly not of the clergy—the common people must do it. When Forster asks in *Cook, the Discoverer*, "Can we not also protect the flower of enlightenment, appreciate and enjoy it?,"[152] the allegorical language may seem to belie the concrete demands he has in mind for shaping the political. Nevertheless, he understands Enlightenment to mean not merely "a more complete, more correct cognition," but a "completion that goes beyond correct application to all events in life."[153] Thus, Forster declares "tolerance and freedom of the press and the unrestricted study of all conditions which are important to man in the name of truth."[154] Enlightenment is a project of communitarization under the provision of a critical public—if necessary, against the prevailing order.

Cook, the Discoverer is thus a fascinating document of the awakening revolutionary consciousness. Forster is not naive; he knows what is at stake in political revolution. Moreover, he is unsure whether radical upheaval will be successful. In a December 20, 1783, letter to Jacobi, Forster writes: "I believe the globe and its inhabitants are in store one day—God knows when!—for a change, *du tout au tout*—and that is indeed the reason why I doubt the perfectly happy success of all reformers. That which they call polluted air will never cease to be until the current state of things is changed—of course, circumstances can sometimes bring about a *change in the weather for a time*, which indeed prevents all from being lost." To be sure, Forster seems to regard the European state as enjoying the calm before a storm, because it can certainly be that "a powerful, wealthy and opulent state in the full enjoyment of its powers already carries within itself the seeds of decay, which is basically an empty word since no society remains viable forever." Be that as it may, revolution remains an inevitable fact for Forster. "This much is certain: the world is taking hasty steps towards the development of a great global event."[155]

The furor of this realization may speak to what Forster was prepared to sweep aside in *Cook, the Discoverer*. The text culminates in a revolutionary apocalypse. Forster had previously only casually outlined his vision for a "great revolution of the globe."[156] This looming apocalypse will grow from just a tentative start into a global event:

> It is a bold thought that five to six hundred million people do not even dream of how seriously and lovingly the philosophy of their brethren is already calculating the means of enlightening them, and that they are no longer far from the moment when a remarkable revolution will take place in their thoughts and actions, when the wise teachings of Europe, per- haps even those of America and the southern lands, will irresistibly com- pel them to renounce their long accustomed slavery, their natural softness and indolence, the desultory processes of their reason, which merely plays with pictures, in short, to show them the errors and deficiencies of their hearts and minds, whether inherited or induced by their climate, in order that they might recognize and accept the truth which renders the Euro- pean independent thinker happy.[157]

In the years leading up to the French Revolution, this might have been taken for daydreaming—or speculation of a historical-philosophical bent, without the added value of *realpolitik*. The broadness of its formulation speaks to the fact that Forster did not yet trust his own words. But Forster is quite the canny diagnostician of his time. Just five years before the great Revolution, he determines that "a new day is dawning among the learned; it is a fermentation of their minds, and it seems, the people will go their own way of thinking."[158]

IV

Views of the Political

The Revolution, 1789–1793

:: :: :: :: ::

In April 1788, Forster takes a position as a university librarian in Mainz. July 14, 1789, marks the start of the French Revolution, with the storming of the Bastille. In 1790 Forster embarks with the young Alexander von Humboldt on a journey that takes them first along the Rhine, from Mainz to Dusseldorf via Cologne, then to England and France via Aachen, Lüttich, Löwen, Brussels, Antwerp, and Amsterdam before Forster returns to Mainz. In September 1792, France abolishes the monarchy and becomes a republic. In the course of the Revolution's battles, Mainz is taken by French troops in October 1792. Forster throws all of his support behind the revolutionary upheaval in Mainz. He becomes president of the Jacobin Club, which is committed to the aims of the French Revolution and strives for abolition of the monarchy. On March 17, 1793, the Rhenish-German National Convention assembles; the next day, Mainz declares the territory spanning from Landau to Bingen a republic. The Rhenish-German National Convention resolves to petition the French National Convention for the accession of Mainz to France. Forster, the convention's elected vice president, is dispatched to Paris to present the appeal with two fellow delegates. While Forster is in Paris, Mainz is besieged by Prussian troops and capitulates on July 23, 1793. The first republic on German soil lasts barely four months.

:: :: :: :: ::

Paris Unrest and the Political Public Sphere

"What do you think about the Revolution in France?" Forster asks in a letter to his father-in-law, Christian Gottlob Heyne, dated July 30, 1789. Just fourteen days had passed since the storming of the Bastille. The news from revolutionary Paris was still gradually making its way into the European consciousness. Forster had no sooner expressed his amazement that never before had "so great a change cost so little blood and devastation"[1] than he was bemoaning the failed harvest, the rising prices of firewood, and the impending bad vintage for wine. Yet the Revolution gets only a few lines. Even Heyne is reticent, tracing the political unrest back to the conditions in France: "But what do you have to say about the Revolution in Paris!" he adds in a mere postscript at the end of a letter to his son-in-law, dated August 2. "Standing armies cannot help, either, once the spirit of liberty has been awakened and the conditions are favorable. Because the latter," Heyne writes, "was the main issue indeed."[2] The Revolution is first and foremost a French event of local concern. It can be attributed to a particular situation that does not necessarily exist in other countries. In a letter from Paris, Wilhelm von Humboldt poses the question, "When will other nations one day begin to follow such an example?"[3] But at this time, such things were still out of the question.

And yet, the unrest in Paris marks a new and final stage in Forster's life and thought. The "political storm" exposes a historical tension that he cannot avoid, nor does he wish to. He, the indefatigable author, translator, essayist, and letter writer, was already lamenting six months before the French Revolution that there was "too much writing, too little action in the world" for him. Although his "somewhat philosophical style" does not lend itself to his becoming "a real demagogue," he does not want to stand by in the face of political unrest: "My little contributions must get made." This is an understatement. Within a few years Forster underwent a radicalization, causing drastic and at times fanatical features to emerge. "One is either for absolute freedom or for absolute tyranny. There is no middle ground," he proclaims, wholeheartedly adopting the Jacobin motto "To live and die as a republican." He emphatically made the Revolution his issue. A year before his death, he looked back and wrote, "I am to be spoken poorly of throughout Germany, passing for the chief instigator of all that is foul in Mainz," that is, the declaration of the Mainz Republic. Forster, the naturalist and world traveler, became the central revolutionary figure in Germany of his time: "which is to say, new, politicized."[4]

The all-important upheaval of Forster's epoch plays out in the last stage of his life: Politics is turned into a public concern. It is no longer negotiated behind closed doors, but rather in a realm of open debate, to which a name was only just being given. In those decades, the English spoke of *public opinion*, the French of *esprit public*, and Forster himself of *public spirit*.[5] In the context of the French Revolution, the concept emerges of an *opinion publique*, or "public opinion," as Forster translates it. Jürgen Habermas emphasizes, "Forster seems to have given currency to *opinion publique*, as '*öffentliche Meinung*' initially in the Western part of Germany, thereby distinguishing between 'public opinion' [*öffentliche Meinung*] and common spirit [*Gemeingeist*]." For Forster, a political public of this kind is nothing less than "the tool of the Revolution, and at the same time, its soul."[6]

The French Revolution thus politicized an entire epoch, since the political was no longer reserved for a few rulers and their administrative apparatus but became a matter of the people. This is to be taken literally: When Forster was traveling with Alexander von Humboldt in 1790, he was surprised to find that "over his bottle of beer, the common man becomes politicized by the rights of humanity and all the new subjects of contemplation, which have finally come circulating to the countryside over the course of a few years." And he recounted a ride on the mail coach from Aachen to Lüttich with eleven other passengers: "There was no end to their conversations about political subjects."[7] By contrast, in 1779, Duke Carl August of Saxony-Weimar-Eisenach appointed Goethe to a privy council or, more precisely, to a "privy legation council," which involved a seat and voting rights in the "privy consilium" that reported to the duke about tasks regarding the administration of the dukedom. The political was still a private matter here, the public deprived of deliberations; it was tangible only in the consequences of the authorities' rulings. During the Revolution, however, Forster sees people "diligently reading newspapers in public houses and coffee houses," as political matters were increasingly being covered in their flourishing pages.[8] Ever since the 1870s, Christian Friedrich Daniel Schubart, just to name one example, was writing and distributing his *Deutsche Chronik* as a "public paper, in which everything that happens in the country should be reported, described and argued." Schubart did not leave out the political; he explicitly reported about proceedings in the country. It was in this vein that he made public the sale of twelve thousand Hessian soldiers to America, for which the Landgrave of Hesse-Kassel received an annual sum of 450,000 talers, while those bartered away "were largely finding their graves in America." To enlighten the public in this way was staggering, as Schubart himself knew: "But there go the hands, thrown up in horror, and calling: Why expose the country's affairs?—As though all of our country's affairs were state secrets." Schubart was imprisoned without trial for his work and confined for ten years, from January 24, 1777, to May 18, 1787, in the Hohenasperg fortress near Ludwigsburg. "Brainwashing in the direct mode still existed," as Habermas notes.[9] Schiller was not the only one to keep Schubart's fate in mind as a cautionary example of political despotism. He was just twenty-two years old when he visited Schubart in prison. Yet it did not dissuade him, a military doctor at the time, from a career as a writer.

Forster worked in the medium of the daily newspaper, too. Like Schubart, he extolled the "freedom of the press" as "divine publicity!" He used it to declare war on despotism in the *Neue Mainzer Zeitung oder der Volksfreund* as of January 1, 1793. Still, there was resistance: "How the times are changing!"[10] The charm of this formula lies in its professed inevitability about the transforming course of time. However, the French Revolution was not a likelihood at first. It came as a surprise and was initially thought to be a conformist attempt at reform, the failure of which grew into a public revolution.

It was looking bad for the French state as the eighteenth century tapered to a close: the financial situation was devastating, with high government debt demanding measures that King Ludwig XVI did not have the sole power to impose. In February 1787 he convened a meeting of the Assem-blée des notables (Assembly of notables)—an expanded council of the king, composed of high-ranking representatives of the nobility, the clergy, and functionaries of the state—which did not lead to the tax increase desired by the king. So, on July 5, 1787, he announced the convening of the Etats generaux (Estates General). This tradition of representation for the three classes, nobility, clergy, and the *tiers état*, or third estate (citizens, tradesmen, farmers, and day laborers), dated back to the Middle Ages. But "representation" is by no means to be understood here as commensurate codetermination. According to careful appraisal, the nobility made up 1 to 4 percent of the total population, but it possessed unparalleled privileges compared to the third estate.[11] Hence, an assembly of the estates was in no way an expression of the will of the people, because their convening entirely depended on the king.

When the estates assembled in 1788, the Crown had more to deal with than just the increasing government debt. Social discontent was on the rise owing to stagnating wages and climbing prices that resulted from a failed harvest. What farmers were expected to deliver under feudalism—in that system, the use of land was provided by landowners in exchange for taxes paid by farmers—was increasingly felt to be burdensome and inequitable. This led to the first unrest.

The inner paralysis of the monarchy now proved to be the real problem, "because," Hans-Ulrich Thamer writes in a pointed assessment, "neither the increasing social tensions within the estates, nor the demands for participation by citizen groups, nor even the spread of a way of thinking aimed at freedom and civil equality, led to the collapse of the *Ancien Régime*, but rather the incapacity for functionality and reform in a moment of growing financial and economic crisis." It was financial weakness and political impotence on the part of the decadent feudal aristocracy, according to Albert Soboul's judgment, that caused all reform to come too late, leading to the spread of conflict across the entire nation.[12]

The discontent of the great majority of the population found its intellectual expression in the pamphlet *Qu'est-ce que le Tiers-État?*, published by Emmanuel Joseph Sièyes in January 1789. In this treatise on the meaning of the Third Estate, Sièyes posed three questions and gave pioneering answers: "1) What is the Third Estate? *Everything.* 2) What has it been until now in the political order? *Nothing.* 3) What does it want to be? *Something.*" In this radical appreciation of the lowest estate, Sièyes formulated a claim to the sovereignty of the people, and it spread like wildfire. In an ingenious move, he explained the third estate to the whole French nation while denying the nobility a place within it: "Such a class, surely, is foreign to the nation because of its *idleness.*"[13]

The Third Estate first found pragmatic expression for its discontents in the *cahier de doléances*, or registers of grievance, which were kept by each district and given to the deputies of the assembly of the estates. A drastic increase in bread prices in 1789 brought things to a head rather concretely in the form of unrest, looting, and uprisings. A tradesman now had to shell out about half his pay for bread—fair prices were out of the question.

The assembly of the Estates General in Versailles starting May 5, 1789, bobbed along with no viable results—until the representatives of the Third Estate took the initiative and formed the National Assembly on June 17, with the aim of giving France a new constitution. Freedom and equality were to be the guiding maxims. The inability of the monarchy to bring about reform caused the ailing Ancien Régime to collapse within a few months.

There were still two powers in the state, though: the reigning king and the National Constituent Assembly (Assemblée nationale constituante). National bankruptcy had not yet been averted. Unrest among the rural population continued to swell, and with it, the "Grande Peur,"[14] or "Great Fear," broke out when peasants began to panic over an alleged famine plot by the upper classes, ushering in the Revolution and causing people to take up arms. Sparks flew in Paris when, on July 14, 1789, eight thousand armed citizens stormed the Bastille, a prison regarded as a symbol of despotism. The guards were murdered, and the impaled head of commanding officer Bernard Jordan de Launay was carried through the streets of Paris. It was an early beacon of revolutionary violence.

Still, the aim of the Revolution was not to abolish the monarchy, but the National Constituent Assembly and the political clubs that had long held the reins. On August 26, 1789, the National Constituent Assembly passed the Declaration of the Rights of Man and Citizen, which formalized in its first article the freedom and equality of all people. This declaration was, according to Soboul, the "Ancien Régime's death certificate."[15]

Georg Forster had firsthand information about all of this. In a year overturned by the events of the Revolution, Wilhelm von Humboldt paid a visit to Forster that September. An eyewitness in Paris, he provided an in-depth report about the new "Parisian, but not paradisiacal, freedom."[16] Forster was still the spectator. But he immediately grasped what made the events so novel. France was "highly curious for the observer," he wrote to Jacobi. "It is an interesting prospect—not *that* they fought, but *how* they fought."[17] There had always been uprisings, unrest, tumult, but this was new: a revolution! A hundred years earlier, the Glorious Revolution had been successful in England, as the Bill of Rights strengthened parliamentarism and broke the absolutism of the kingdom, but that was history. The Paris unrest pulled Forster into the maelstrom of the political present. "I have hitherto escaped the cycles of great revolution. Now I find myself smack dab in the middle," he said just a few years later.[18] But he was not quite there yet.

Historical Signs of the New World: Revolution

Five years before the unrest in Paris, Forster spoke of "a rather recent revolution in my thinking, which I hope will contribute a great deal to my contentment in the future. I still have a good portion of fervor to unloose, and thank God this will occur before my thirtieth year is through." Forster was alluding to his renunciation of the Rosicrucians, with whom he had been staunchly affiliated for some time. Before 1789 "revolution" was still an apolitical term that denoted fundamental change. Even Kant refers to a shift at that time in his philosophy of *Erkenntnis*, or cognition, as being a "revolution in this sphere of thought."[19] After the political events in France, he called for a "gradual *Reform*," with a view to the moral perfection of man, "effected through a *revolution* in the disposition of the human being."[20]

The politically charged term "revolution," as it is understood and used to this day, has, strictly speaking, "become possible only since the French Revolution," according to Reinhard Koselleck, and it is only through the Revolution that it has received "consecration through necessity."[21] Throughout history there have always been political confrontations, but revolutions are utterly modern. Neither antiquity nor the Middle Ages knew revolution as the overthrowing of a political system. Plato had already run through the advantages and disadvantages of various possible constitutions of the state in the eighth book of his *Politeia*, but that was theory. It did not result, as a practice, in the revolution of the state.

The word *revolutio* is first found in Christian literature of late antiquity, where it denoted, among other things, the rolling over or rolling aside of the stone that sealed Christ's tomb. It also denoted the natural occurrence of the rotation of celestial bodies around the Earth, as in the title of Nicholas Copernicus's seminal work *De revolutionibus orbium coelestium* from 1543.

Only on the eve of the French Revolution did *revolutio* begin to take on a political meaning. If Copernicus could speak of the Earth's revolution around the sun, so, too, could Diderot and d'Alembert's *Encyclopédie* designate as *révolution* the British Glorious Revolution of 1688–89, when William of Orange seized his father-in-law Jacob Stuart's position. "King Jacob's poor administration . . . caused the *Revolution* to necessarily materialize & made it possible," Louis de Jaucourt writes in the *Encyclopédie*,[22] in explaining how the central political position underwent a change.

Revolutions belong to the distinctive signature of political modernity, having inscribed themselves into the political consciousness of the Western world. A look at an American one-dollar bill refers a person to its revolutionary constitutive act. The unfinished pyramid on its back symbolizes the building of American society, with the year 1776 inscribed at its base as a reminder of the date of the Declaration of Independence—the pinnacle of the American Revolution. With this revolutionary act, a "new order of the ages" had begun—novus ordo seclorum—as a banner beneath the pyramid reminds us. It was not the first revolution to attempt a reconfiguration of political structures through a historically powerful break, nor was it the last to lend world historical concision to social developments, as well as developments in the history of ideas. The French Revolution of 1789 changed not just one country, but all of Europe—despite all efforts to restore reactionary political systems—by drafting a new ground plan for political modernity.

Revolutions not only cofounded political modernity but were themselves instruments of the political. They are an originally modern means of compelling modernity. Anyone who wants to understand political modernity must understand its revolutions. This is something that Georg Forster could already sense as he stood at the threshold, with one foot in the old world and one foot in the new.

For him it came down to the necessity of political change. Rebels and agitators disturb the current order, whereas revolutionaries leverage a new order. They stand at the service of historical necessity, seeing themselves as its executors. The modern revolution thus legitimates not just a sweeping overthrow, but the force required to achieve it. The means of force is permitted, if it cannot be avoided, when attaining a higher purpose. In the revolutionary drama about the hegemony of the old or new order as it is performed on the stage of history, violence appears as the driving force. The die is cast, Forster is thus able to exclaim in 1792, entirely in line with this revolutionary thinking: "Jacta est alea! Let us no longer speak of principles. A public summons had been issued for the rights of the strongest. Let us see who it will be."[23]

To Forster, a revolution is a historical event, which one does not initiate so much as observe, accompany, and encourage as necessary. Although one must take sides, each individual has only a small stake in the entanglement. In January 1789, six months before the storming of the Bastille, Forster was already making sense of the dynamics at work in the Revolution, which he compared to a moving ball: "Countless hands come in contact with the ball, as it is thrown, pushed, grazed, touched, and all of these different small and large impulsions propel it forward." There are forces at work in a revolution that no single person can quell. The example of the ball is not quite as harmless as it may first appear. Three years later Forster returned to it to illustrate the hubris of the revolutionaries: "They dared to push away the destiny of the State like a ball whose direction they determine. Thereupon, the ball grew too large for them, and they had to let it roll where fortune wanted."[24] The revolution, and not the revolutionary, specifies the direction. The revolutionary merely rides on the back of the tiger.

Even cautious minds were impressed by the idea of the revolution's unstoppable, dynamic progress. For Kant, though, the French Revolution was a "historical sign." Although human progress cannot be directly experienced, "the *tendency* of the human race viewed in its entirety" is authenticated through events of this kind.[25] The enthusiastic and selfless public advocacy for the French Revolution's issues by its sympathizers—their partisanship risking an adverse effect—speaks for the system, in which the events of the Revolution are morally inherent.

In the great Revolution, Forster, like Kant, sees the breakthrough of an inevitable human emancipation movement. There is notable agreement across two testimonies on this. On July 12, 1791, Forster writes in a letter to Heyne: "The political world runs exactly as one can expect it to after what has preceded. People have been taught to be free immature beings, brought up to be formed into mature beings, and shamefully abused, made dumb and blind, while the rulers freely grant themselves intelligence and satisfy their passions. Is it any wonder that outbursts of this feeling, now that it is finally awakened, cannot be entirely pure and unblemished?"[26] This goes beyond a cautiously framed apology for the use of force in revolution, revealing the continuity of Forster's thinking—spanning the period from his voyage around the world to his revolutionary years in Mainz. In Tahiti, Forster first considered the tendency toward decadence in accounting for the privileged class's "voracious appetite," which he now compared to the European nobility's unique position. On the shores of the South Seas, Forster was already conceiving of a historical dynamism, which held true for the current revolutionary outbreaks. "At last," Forster writes in *A Voyage round the World*, "the common people will perceive these grievances and the causes which produced them; and a proper sense of the general rights of mankind awaking in them, will bring on a revolution."[27]

Certainly Forster joined the choir of the Enlightenment from time to time and declared the revolution a triumph of reason. It is nice to see, he concludes, "what philosophy has matured in the mind and then brought to the State."[28] Praising reason in this manner, though, is an intellectual conceit on Forster's part, derived from the occasional opportunism of the Zeitgeist. Forster in no way conceives of revolution as the result of reason, but as the consequence of an "awakening feeling," as the two passages demonstrate. "Revolution" does not map onto "reason."

The French Revolution may have been a complete surprise to his contemporaries, unexpected and unique, but ever since his time on the shores of Tahiti, Forster had been preparing a history in which revolutions are "the natural circle of human affairs." It is unavoidable, he writes, upon completing *Cook, the Discoverer* in 1787, that the "relationships between the old world and the new" should be given "an entirely new aspect . . . by means of these new revolutions."[29] Forster was prepared. But not by those philosophical writings that, only in in hindsight, could be said to be intellectual precursors; rather, by his own immediate experience and feeling—when taken together, by what he would have called political "realism." Perhaps that is why he was more readily able to read revolutions as historical signs. What is certain, however, is that when he saw his understanding of history confirmed, he styled himself all the more resolutely as a German revolutionary.

Political Views of the Lower Rhine

Forster sees the French Revolution's promise of freedom and equality as an echo of his experiences voyaging around the world. It is unsurprising that he did not hesitate for long before siding with those experiences. For months before the Paris uprising he had moaned that there was "too much writing, too little action in the world,"[30] but now he was attempting to become a man of action—by reaching for his quill.

And so, once again, Forster took to traveling. He was sorely in need of a change of scene. In April 1788 he took a position as a university librarian in Mainz. The year before, he had received an offer to guide a Russian expedition to the South Seas—he was eager to leave dreary Vilna behind, and the cost of dissolving his contract would be settled by the Russian Admiralty. No sooner had he arrived in Göttingen, though, than the expedition was thwarted by the Russian-Turkish War, leaving Forster with no choice but to take a librarian job in Mainz. At least it offered his family an income: "So we will be on solid footing again," Forster wrote his wife.[31] He and his family moved into several rooms of a faculty guesthouse, which to this day has been preserved in Mainz. Forster's second daughter, Clara, was born there on November 21, 1789.

A city of about thirty thousand inhabitants, Mainz was open-minded to the spirit of the Enlightenment—or so it seemed. The university decorated itself with professors of the Enlightenment, for example the Kantian Anton Joseph Dorsch and the natural law theorist Andreas Joseph Hofmann. In Mainz Forster met bright minds, among them Adam Lux, who was promoted to *Doktor* in 1784 by the university for his philosophical work on enthusiasm. Lux struggled along as a farmer and a tutor, and—how else could it be?—as an ardent revolutionary who eventually accompanied Forster to Paris. Compared to Vilna, not too shabby.

The appearance was deceiving in many respects, though. The library, for which Forster was responsible, was in pitiful condition. None of the rooms were adequate. Of the approximately fifty thousand volumes, the vast majority were duplicates. The primary theological literature was of little interest to the Enlightened intellect. "The usefulness of this library, for the purposes of teachers and pupils, is practically nil," Forster wrote his father-in-law.[32] Despite his efforts, Forster was unsuccessful in persuading the electors to furnish better resources.

He returned to writing and to the tiring work of translating. It was not long before he found himself lacking in stimulating impressions, his thoughts becoming idle. He wrote in the winter of 1789:

All winter long I must compile and translate! My head is empty, I have nothing more of my own to say of the world. To travel to Italy, or England, or Spain, or even farther, where there are only new things to be seen! Because in the end, one has nothing more than what pours through these two small openings of the pupils and excites the vibrations of the brain! We do not take in the world and its essence in any other way. The poor twenty-four *signs* do not suffice; the *presence* of things and their immediate *influence* is something utterly different. I will soon turn thirty-five years old. The best—by far the best—half of my life is behind me; and has passed me by in vain.

What's more, his poor physical condition was an ongoing issue. Forster sought a change in scene, since travel represented for him the "great incomparable source of the most certain instruction by one's own senses," and his "health require[d] a bit of agitating."[33]

Forster took a vacation, setting off along the Rhine on March 25, where he passed through Koblenz, Cologne, and Dusseldorf. He visited Aachen, Lüttich, Löwen, Brussels, Lille, Dünkirchen, Antwerp, Rotterdam, The Hague, and Amsterdam. He crossed the English Channel, whereupon he of all people—he who sailed the oceans and once called the North Sea "a big puddle"—became seasick. He stayed in London and toured the west and north of England, before departing Dover for Calais. He spent five days in Paris before returning to Mainz on July 11, 1790. His travel companion was a "quite congenial, well-behaved, spirited, knowledgeable young man," Alexander von Humboldt.[34] Throughout the whole trip, Forster kept a journal and wrote letters to his wife, as a way of recording his immediate impressions. He referred to them later while working on a book that touched the nerve of the Zeitgeist. His travelogue *Ansichten vom Niederrhein, von Brabant, Flandern, Holland, England und Frankreich, im April, Mai, und Junius 1790* (Views of the Lower Rhine, from Brabant, Flanders, Holland, England, and France, in April, May and June of 1790) was geared toward the "general public" and "the broad mass of readers."[35] As such, he chose a stylistic device that he called the "light tone of a letter," so that the book would "read well," and he insisted to his publisher that the text should be "printed in elegant typography."[36] Public success, as Forster knew all too well, is no accident.

Schubart's *Deutsche Chronik* praised the first volume of *Views of the Lower Rhine* as one of the most superb travelogues in all of German literature, as did Lichtenberg, who deemed the book "one of the first works in our language." Even Goethe expressed praise, in spite of distancing himself from Forster's political stance. The book pleased not only him, "but everyone who has read it," he told Forster in a letter.[37]

Perhaps not everyone. Forster's father-in-law, Christian Gottlob Heyne, whom Forster was very close to, felt the explosive potential of the book even before its publication. "You are touting the news of your travels. I felt afraid reading it," he writes in August 1790. "You are walking on coals," he cautioned. After its publication, he needed a while to overcome his "initial aversion," as there were indeed "wonderful parts in it."[38]

Views of the Lower Rhine was not an innocuous account of entertaining impressions from Forster's travels, but an intricate web of present-tense descriptions from varying perspectives. Forster had been holding out the prospect of the manuscript to Johann Karl Philipp Spener, his publisher of many years, with the promise that he would "draw up [his] notes" about the regions he would travel through, their natural resources, their current political situations; about their people, their character, their physical attributes; about the works of art that he encountered "here and there,"[39] the factories that he visited—basically everything that struck him "as noteworthy." Even this list alone is noteworthy. Forster does not provide a travel guide with practical tips for readers, oriented toward the popular points of interest. Only occasionally does he touch on what one might expect, such as the museums he visited, whose works of art he describes. He finds such details superfluous, because "one no longer travels to Europe as they would to *Terra Incognita*. There is virtually no small town or country that does not have boring topography to exhibit; one knows in advance what is to be found and seen there."[40] Forster has something else in mind.

Views of the Lower Rhine is an artful book. Even its title is studied and ponderous. In its complete version, it is a bit prolix; only in the abridged form is it memorable. Such elaborate titles were reserved for accounts of expeditions. If explorers wanted to emphasize the exceptional length of their voyage, they would fully enumerate the years in the title. Accordingly, the title of Cook's account of his second circumnavigation is *A Voyage towards the South Pole and Round the World, Performed in His Majesty's Ships the Resolution and Adventure, in the Years 1772, 1773, 1774, and 1775*. Alexander von Humboldt follows suit, with his *Voyage aux régions équinoxiales du Nouveau Continent, fait en 1799, 1800, 1801, 1802, 1803 et 1804*. Forster's enterprise was not of comparable duration, yet he specifically mentions the months of his travels in the title: April, May, and June. To compensate, as it were, he lists the five countries he travels to—Brabant, Flanders, Holland, England, and France—even though he was in France for only a few days. No exotic destinations, no utopian promises or exploration of the world. From the publisher's perspective, Forster's book title was problematic for a popular, nonscientific work—Spener ultimately did not publish the manuscript, but Christian Friedrich Voss was willing to do so. Out of gratitude to Forster, Humboldt named his most popular book simply *Views of Nature*, furtively tacking on *with scientific annotations*. It is because of Forster that his work is included among the series of expedition accounts that open up a world to the reader. Since discovering England or France was out of the question, Forster's travelogue introduces the reader to political change in Europe instead. *Views of the Lower Rhine* is an expedition into the new world of incipient republican politics. Unfortunately, Forster never wrote the third volume, which was to deal with England and France, specifically Paris. He no longer felt "any urge to do so."[41] But the capital of the Revolution clearly serves as the vanishing point of his journey.

But that's not all. Forster's title contains a double meaning that is not immediately apparent. It speaks of "views," and the book's late success makes it easy to forget that there is something unusual to be seen. By comparison, Karl Philipp Moritz titled his 1785 travelogue about his voyage to England *Journey by a German in England in the year 1782: In Letters to Herr Oberkonsistorialrath Gedike,* whereas Goethe simply published under the title *Italian Journey.* In Forster's case, the dual meanings of his title signify that he does seek to reproduce through the travelogue his immediate impressions of the journey, namely, the "views" he takes in; however, rather than leave the matter to purely subjective impressions, he also wants to present the reflected, intellectually suffused "views" of the author.[42] The title should, according to Gerhard Steiner, "give expression to the dynamic relationship between what is ascertainable by the senses and by reflection." Forster wants to "use views to guide insight."[43]

This was an ambitious design, and Forster admitted to Schiller that insofar as the letters he wrote on the journey presented his *Views,* they were entirely "too tediously and cumbersomely wrought" to be able to pass as letters in the proper sense of the word. But that, too, is not all. The prose in the first volume of *Views of the Lower Rhine* is over-written; Forster wrote the second part in a "less taut style." And he prefaced all of this with a dedication that ended up "a bit dark"—Ulrich Enzensberger aptly called it "so stilted that it generally causes perplexed head-shaking and Forster tried to explain it to his publisher Voss." The dedication itself, though, is not as important as Forster's willingness to risk not being understood by his readership.[44] At least not readily.

This is because Forster's *Views of the Lower Rhine* has the character of a political pamphlet. Given the familiar geographical and cultural terrain through which Forster moves, what matters is only "what one sees, and how one can awaken in others a more or less vivid mental image of objects from received sensory impressions, or better yet, activate their powers of imagination so that they can create for themselves a verbal phantom that has a few traits of the real."[45] If what is important is not *what* but rather *how* one sees, *Views of the Lower Rhine* offers the reader a chance to try out an unaccustomed perspective: Forster applies now to European states the cultural, political, and ethnological point of view that he honed during the voyage around the world through his encounters with exotic peoples.

The reader's perspective should be developed, refined, shaped, awakened. Forster had already inscribed revolutionary political doctrine into *Cook, the Discoverer*, by presenting Cook as the personification of the Enlightenment's civilizing project. Cook's discoveries around the earth were going to have "deep roots and long have a decisive influence on the activities on mankind." Abiding by his botanical metaphor for progress as a process of growth, Forster now explains, with regard to *Views of the Lower Rhine*, that he must "tread cautiously," because he wants to "scatter the good seeds such that bystanders do not notice them being sown."[46]

Readers did notice the seeds of republican freedom, though, and that explains the lukewarm response by many review outlets. The first volume nearly got pulled after its publication. "I know it has its flaws, its faults," Forster concedes, "but it is not a bad book." Heyne advised Forster to trim the metaphysical and political *raisonnements* in the second volume, so that he wouldn't be "lacking in general acclaim from the public, and it must make a sensation, even among those, who merely want to be amused: and they make up the greatest lot." Forster did not like following this advice, because the book's aim was to overcome precisely that resistance which Forster presumed the reader would have. To that end, he proceeded cautiously to calm nervous minds by pointing out that "nowhere—therefore, not here, either" were European people, with the exception of France, "ripe for lasting revolution, neither the church nor the political constitution." Previously he had sought to calm the frightened: "Incidentally—as consolation for all poor sinners upon and at the foot of the throne—one thousand years is perhaps the minimum length of time for a revolution such as this."[47]

Forster's attempt at politically educating his readers is not grounded in arguments, but primarily in strong images taken from his letters. One of the strongest can be found—not merely out of chronological necessity—right at the beginning of his travelogue. Forster's description of his visit with Alexander von Humboldt to the Ehrenbreitstein fortress, near Koblenz, illustrates the direction in which Forster steers the reader's attention:

> We scale the Ehrenbreitstein. Not the unimportant treasure of this fortress; not the griffin bird, that immense cannon, which is supposed to shoot a cannon ball of a hundred and sixty pounds all the way to Andernach, but has never done so; not all the mortars, howitzers, culverins, twelve- and twenty-four pounders, long barrels, canister-shot bushings, hulled wheat, and whatever else is to be marveled at in the armory or on

the ramparts; not the expansive view from the highest mountain peak,
where Koblenz lies map-like, with the Rhine and the Mosel, below one's
feet—none of this could make up for the abominable impression made on
me by the captives there, rattling their chains and holding their spoons out
between the bars of their smoky window grilles in order to take alms from
sympathetic passersby.[48]

In a single sentence, Forster shifts the focus from attractions—which he
only mentions to make them appear not worth mentioning—to the pitiful
sight of the prisoners.

He does not inquire about their crimes, nor the lawfulness of their punishments—are they dangerous criminals? As a result, the image conjured by
him remains indefinite and inured to metaphor. Enlightened readers would
have had Rousseau's dictum in their ears: "Man was born free, and everywhere he is in chains." It is precisely those chains of oppression that the
reader hears rattling in Forster's words now. The disparity is blatantly clear:
here, the military insignia as an expression of strength and support for the
despotic power of the state; there, sympathy for the suffering creature. This
first view from the journey already elucidates what is important to Forster
and what is not. Lichtenberg immediately recognized the secret blueprint of
Views of the Lower Rhine: "I read it as a book about people."[49]

Reflection follows the description of the pitiful prisoners. "Would it not
be fair—my heart was struck by this—that whoever sentences people to
prison should have to hear their cries, how their laments pierce the heavens,
with his own ears at least one day a year, so that he should be convinced
of the lawfulness of his judgments not according to the dead letter of the
law, but according to his own feeling and living conscience?"[50] The political
should be guided by concrete experience, immediate experience, feeling—
not caprice and defunct laws.

The image of the prisoners in Ehrenbreitstein did not fail in its desired
effect. The *Deutsche Chronik* mentioned it in its review and spoke of the
"gentle, humane heart" of the author. Yet sympathy was not the only feeling that Forster sought to evoke. When he and Humboldt went to Aachen,
their itinerary dutifully included the cathedral. "I have visited the cathedral," Forster writes, a sober start to his account, to segue to the famous
Royal Aachen Throne. He offers a disappointing sight. "The throne on
which so many German Kaisers have been crowned since Charlemagne's
time is made of rather plain white marble and has such an undignified shape

that you might mistake it for a parody of all the world's thrones."[51] Even here, the description's metaphorical excess is striking: the shabbiness of the throne extends to monarchy as a form of government.

Here, too, Forster's reflective, reasoned view of the political immediately follows his rendering of sensory impressions:

> The history of the last century has passed my memory right by. What Vienna, Regenspurg, and Wetzlar have presented as very different ideas about the essential components of the Reich's constitution; how the Kaiser's role has been gradually limited, through all its metamorphoses, up till its current form, where only the shadow of former dynastic power remains; how the numerous free estates, now under the irresistible superiority of an all-wealthy few among them, content themselves only in the name of freedom now and must approve the legislative will of these few: all of this fills me with the crushing conviction of how little the larger course of global events reveals about the arbitrariness of peoples' fates, the dignity of sentient beings; and how the happiness and welfare of millions who crawl around the world constantly remains dependent upon dead letters, upon slavish adherence to ceremonies that have become devoid of meaning, upon the baseness which gives importance to empty minds—and consists in no way of their own power and deeds![52]

These were strong words.

And Forster does not shy away from openly contemplating the "instability of the constitution"—which means pondering only its possible change. In one of his preliminary letters to his wife for *Views of the Lower Rhine*, in which Forster recounts his visit to the Aachen cathedral, he describes his opinion of the throne even less reservedly: "If a German prince wanting to be Kaiser were to stand before the throne and see it the way I do, perhaps then there would be an end to this tangled mess of inane formalities." He continues, "If our great princes were to become wise enough to deride the feigned dignity from which they alone benefit, and if their positions were to be taken up by lesser men, who have been yearning to step into them, the Kaiser's name would be left to the mockery of children, and we could finally became awake to the fact that this frolicking in the name of the German Reich in Regensburg, in Wezlar, and elsewhere, is not merely contemptible to other nations, but is not suitable to the dignity of sentient people, either."[53] Germany would then receive a constitution that did not consist of mere words, "foolish and costly ceremonies"; rather, it would be powerful and intentional. These are almost seditious watchwords—and as such, Forster himself did not dare to publish them.

He had a feel for gestures and their hidden effects, though. He was offered something that remains forbidden, even to this day, to visitors to the Aachen cathedral: to sit on the throne. "So profusely was I invited to sit upon it that I did not feel the slightest temptation to do so," he writes in a letter to his wife. But here is where the account deviates in *Views of the Lower Rhine*—not by much, but also not insignificantly. "So profusely did the guide implore us to sit upon it"—Forster does not refer to himself in the singular—"yet, I did not feel the slightest temptation to do so."[54] By taking the invitation to ascend the throne and extending it beyond himself to an indeterminate "us," while at the same time reflecting his refusal as an individual, he presents his conduct as a model for others. This dramatized gesture of not ascending the throne illustrates the distance gained over handed-down tradition more strongly than any argument could. Whereas Forster was the lone addressee in the version from his letter, in *Views of the Lower Rhine* his refusal extends to an indeterminate collective, because Forster does not mention that another visitor, namely Humboldt, was present to accept the cathedral guide's invitation. The shift from "I" to "us" gives form to Forster's narrative aims, prompting his readers to imagine how they would react to such an invitation.

Forster repeatedly employs stark contrast as a stylistic device to paint tradition as stale and the dawn of the future as full of promise. Yet he appears deeply impressed by the Cologne Cathedral, that "magnificent temple." The choir, which had not been completed until Forster's day, possessed a "majestic simplicity exceeding all imagination." The columns grew upward "like trees in an ancient forest."[55] At the same time, it was a glimpse into the past, the "darkness waking an eerily primeval picture in the empty, forlorn arches, which our footsteps echo through, between the graves of electors, bishops, and knights, who lie here hewn in stone."[56] Religion is portrayed as a relic of the past, venerable and obsolescent. The Cologne Cathedral is as empty as the clergy's promise of salvation.

Religion is in decline. Forster recounts a visit to a Trappist monastery near Düsseldorf. Though the monks were ordered to silence, the first monk they meet "speaks with us at once," disregarding the vow of silence.[57] The monks' inactive lifestyle has caused them to become fat and sluggish. In the religious atmosphere, life slackens. It is no coincidence that Forster concludes his brief description of the clerics' scandalous conditions with a quote from Goethe's *Prometheus*. Jacobi had given him a copy of the ode in November 1778. Now it serves as a counter to the decadence of the monks: "Each chooses that which serves him; I know that this existence and this end have no appeal for one who has known the better fate of man: 'to suffer pain, to weep, to feel pleasure and joy.'"[58] Forster thought of himself as also an inhabitant of the Promethean world, borne there by a self-awareness that seeks to get along without religious comfort.

With religion on the wane, the authority of the clerics becomes void. A brief sketch that Forster did of the Cardinal of Mecheln sufficed to convey in just a few words the hollowness of ecclesiastical dignitaries. Forster met him at a church in Brussels, "dressed in a long scarlet robe and cape, with a red cap on his wig; a man of quite considerable stature and age, with a soft, loose, fleshy face. He knelt behind the large altar and prayed, inspecting his rings in the process, which poked out of his sleeves." In his private correspondence, Forster confessed to finding the Roman Catholic religion to be "abhorrent above all others, because of its despotic spirit and its intolerance." To his friend Soemmerring he wrote: "We make fools of priests and monks,"[59] and thus he sets the tone with which he exposes the cardinal as a ridiculous figure, whose outward ceremonies are as insipid as the feigned dignity of political princes and kings. With just a few images, Forster portrays the dignity of the first and second estates as having faded, their authority eroding with historical progress and their claim to a principal function in society becoming porous. The elegance of Forster's approach resides not in delivering an argument to the reader that is so convincing as to compel agreement, but in delivering images, or "views," for the reader to interpret. They are clear enough to steer the reader in the desired direction.

Forster, himself a Protestant, predicts no future for the "despotism of priests,"[60] as religion will thwart the progress of humanity. "Spiritual and oligarchical pressure," he cites as his evidence, have "banished hard work from the walls of Aachen."[61] A mere hour away, in contrast, amid liberal and tolerant conditions, Dutch factories were flourishing. This observation is not without political overtones. "How small and base," Forster writes, his reflections following his description of the factories, "does every despot appear, who trembles before the enlightenment of his underlings, when compared to the private citizen, the manufacturer of a free State, who establishes his prosperity on the prosperity of his fellow citizens and on their more perfect understanding!"[62]

To encourage the reader to imagine a strong picture of wealth in a society no longer led by the nose, Forster describes the fruits of an impressive achievement that has been "secured by trade" and the "concentrated powers" of a nation. In the Amsterdam harbor, he witnesses the launch of the frigate *Triton*. Once it became waterborne, "the crowd of daredevils, whooping and rejoicing high above us, sailed away on the new *Triton*; they waved their hats, their voices drowned out by a loud cry of jubilation from the land."[63] It was a triumph of national unity and human ability.

Had Forster completed the third volume of *Views of the Lower Rhine*, he likely would have gotten around to Herschel's telescope in Slough—as a counterpart to the scene in Amsterdam—which he visited during the English leg of his journey. A marvel of technology that the astronomer Friedrich Wilhelm Herschel had built, it was of such immense scale that it had to be moved on casters and wheels, and its magnification was so powerful that viewing the moon through the telescope could blind an observer as severely as looking at the sun.[64] As a symbol of the scientific Enlightenment, this technical masterpiece epitomized the triumph of an epoch that no longer wished to be censured by religion for its view of the world.

Smoke and mirrors can be detected at some points in the many stand-alone descriptions that Forster provides in *Views of the Lower Rhine*, perhaps to distract from the fact that his account does not trace a purely geographical journey. The journey that Forster describes is one that takes place not only in space, but above all in time. Willing to leave the past behind, Forster adopts the role of a guide who is in close contact with the future. Insofar as *Views* is a work "where it comes down to imagination and *Raisonnement*,"[65] it deals less with what is real than with what is possible.[66]

As a result, Forster's role as author changes. From a retrospective writer of travelogues emerges a guide to the imminent future of social life, a guide ultimately concerned with effects. Friedrich Schlegel shrewdly captured this and clearly named it. In his essay "Georg Forster: Fragment einer Charakteristik der deutschen Klassiker" (Fragment of a characterization of the German classics), written in 1797, three years after Forster's death, he seeks to characterize Forster's political writings from a nascent distance. He assigns him the role of the "brilliant *societal writer*."[67] This carries with it the task of "exciting, forming, and uniting anew all essential human endowments,"[68] including the endowment for politics, too, and not just those that are culturally valuable beyond government enterprise. *Views of the Lower Rhine* is a sociopolitical work, whose aim is for the reader to join the author at his observation point for the course of upcoming change. The book does not incite revolution, nor it does demand republicanism. But it does generate a historical plausibility for one as much as for the other. Forster places himself in the service of a liberally oriented mobilization of society that seeks to overcome despotic encrustation for the benefit of republican codetermination.[69]

As Schlegel aptly notes, the descriptive charm of *Views of the Lower Rhine*, as with Forster's other political writings, is derived from its being "tremendously elevated by its scientific veneer." As though he were on a voyage of discovery and were faced with unfamiliar peoples, Forster reports on the conditions in the countries he travels to and extracts generalized conclusions from particular cases. To Schlegel, Forster is "not a true artist of reason"—a judgment Forster would have taken as a compliment, since immediate experience should be the starting point for all political argument. Forster does not conceive of any utopias, or even ideals, as being learnable from philosophy books. To him, ideals are merely "creations of the intellect and too delicately woven to be suitable to reality. That which has practical application must be formed from coarser material, more substantive if you like, but for that reason, all the more natural and more human."[70]

Schlegel concludes by accurately naming the two motives that drive Forster's sociopolitical writing. In *Views of the Lower Rhine*, Forster's political thought presumes the "unwavering *necessity* of the laws of *nature* and the indestructible capacity for human perfectibility."[71] It is between these two poles that Forster's views emit flying sparks.

A commitment to the political, and to not standing in the way of human development, became popular in Forster's time. Republicanism aligned with the ideal of self-determination, according to Forster, in its respect "for man's individual character and an unfettered start, instead of being based on the false tenet of despotism, in which man is created purely for the State and is regarded as a gear in the machine that moves as a single entity." Even Schiller, in *The Aesthetic Education of Man*, attempted to lay the groundwork for a political freedom that sought to protect the individual from the machinery of power. But Forster's relationship to the Enlightenment is cryptic. When he speaks of the "spontaneous combustion of reason in an entire people,"[72] self-determination by the individual is placed just as much in question as the sovereignty of reason. The image is chosen with precision: spontaneous combustion is an occurrence, not a decision. The flip side of the Enlightenment—its dark side, as it were—is the heteronomy of reason through natural forces. Reason is a tool within the whole of nature, not an autonomous source for the organization of the world.

Forster remains faithful to his fundamental premise of a physical anthropology, because he views reason as the subordinate organ of all action, even revolution. It is precisely in revolutionary disjuncture that "feeling decides instantaneously, even before reason can disentangle itself from the chaos of conflicting systems." Thus, no one should come to him with the "wretched platitude which so many apostles of despotism now bandy about, and which, much to my disgust, I have heard repeated by parrots: that the Enlightenment is to blame for political revolutions."[73] The reasons for revolution are much deeper. Revolutionary overthrow is neither an expression of the sovereign will of the people nor of enlightened reason, but of nature, which determines humanity.

Revolutions spring forth from the "powerful, dark drive" of their agents, not from the clarity of reason in the light of its insights. Revolutions have their causes, but people are not in command of them. During his time in Tahiti, Forster had already said, "A proper sense of the general rights of mankind awaking in them will bring on a revolution." And in *Views of the Lower Rhine*, too, Forster understands grievances to be the "indelible seeds of a new revolution, which will take a hundred years perhaps to germinate." In the "tempests of the moral world," humanity is the driver. Revolutions are natural events, in which reason may have its share, but not more. Forster's remarks cannot be understood otherwise: what revolutions create in a state is "entirely independent of each degree of insight by the people revolting."[74]

Forster means to be able to cite an apt example of his theory. In a letter he mentions his intention "to speak of Lüttich, too . . . with some caution" in *Views of the Lower Rhine*.[75] This was a delicate nod. The economic situation in that city did not look good; the cost of living had starkly risen as it had in France, and in 1789 sparks from the great Revolution flew as far as the Prince-Bishopric of Liège: the citizens deposed the old magistrate and chose a new one for themselves. The prince-bishop had to flee. On a smaller scale, Lüttich replicated the Paris revolution.

A model for other cities and nations? Forster assuages such expectations. As a rule, "a forcible dissolution of government is unthinkable, and one feels compelled to aim all efforts solely at halting individual abuses, at the correction of individual mistakes that have large-scale effects and ruin everything. Perhaps, in most cases, it is truly advisable," Forster says, moderately, "to better an old flawed constitution than to organize a completely new one—and put the whole thing at risk through fermentation, which is unavoidable. As with the introduction of all new things, something different than what was hoped for will be gained, or holes and deficiencies will now be exposed, which will perhaps cause greater calamity than what one first sought to redress." A change in constitution is highly risky, akin to "changing the wheel on a moving carriage," as Schiller put it. Thus, Forster does not call for revolution.[76] When, however, a contract violates morality, it becomes void, "and a state constitution does not for a moment have a legal right to exist, if it robs its members of the possibility for moral perfection."[77] The revolution occurs on its own, when the conditions necessitate it.

Forster's view of the events in Lüttich must have unsettled his readers. He did not always find approval by referencing the right of liberty, the triumph of reason, or the history of progress, but he was moving in the riverbed of the Zeitgeist. This would have been forgivable, but this is not how Forster thinks. And he is honest enough not to disguise his view of things. In a passage that does not spare him or his reader, Forster equates Lüttich's history with the course of world history. In his sketch he defies all Enlightenment-era utopias and ideologies of progress. In doing so, he does not deny the dawn of the new world, but he identifies the impetus for historical change in a way that was unheard of. It is understandable why his comments might have taken his father-in-law's breath away. "Violence," Forster begins,

> not by blandly persuasive reason, but by physical dominance, brought forth all change to this small state, as to every other, going as far back as the dark Middle Ages and continuing to unfold before our eyes now. Violence established peace in 1316, despotism in 1684, and liberated the people again in 1789; violence shall support the rulings in Wetzlar; and it is precisely violence, not the excellence and inner righteousness of the matter, which will perhaps secure the constitution in Lüttich. That is the course of world events. Nothing takes place the way the drawn-up rules of reason would have it take place a priori,[78]

which is to say, independent of experience. The entire magic of civilization and the sheen of political institutions result from violent altercations: "Battles precede treaties."[79]

This was not a new idea. Thomas Hobbes had already explored it in *Leviathan*, one of the founding texts of political modernity. Before there was a state, there was man in his primitive state, which Hobbes conceives as being a war of all against all—man is wolf to man. But for Hobbes, it begins with violence, and nationalization consists of overcoming precisely that civil-war-like violence by centering the state's authority upon violence. Pacification is the objective. Forster, by contrast, explains violence as the ongoing and driving force of history.

The observer of nature is better attuned to this. It is more than a "scientific veneer," as Schlegel thought, when Forster first introduces the concept of revolution in *Views of the Lower Rhine* by observing nature: he and Humboldt are interested in basalt, a type of stone whose formation was still disputed at the time, but which had already been linked to volcanism. On their journey along the Rhine, they passed through a former volcanic region stretching from Boppard and Koblenz to Andernach and Bonn, where, as naturalists, they contemplated "the traces of transformations past and great decisive natural events." This sounds harmless, but it is not. For Forster—as for Goethe—there is a profound political aspect to contemplating nature. Forster vividly recalls a panorama of erupting volcanoes for the reader: "Volcanoes steaming and smoldering; molten streams of lava flowing into the sea which covered all of these lands back then, where it would suddenly cool, forming rugged columns; burned out rocks, ashes, and coal flying through the air, and falling down in layers." That was not merely imagined. On Tanna, an island in the New Hebrides, Forster had been able to observe an active volcano. Every five minutes, from the crater came "very violent claps of thunder, and a rumbling noise continued for half a minute together. The whole air was filled with smoky particles and with ashes." Each time the volcano erupted, "its fires afforded us a most pleasing and magnificent sight." He even came quite close to the volcano, where "the earth was so hot, that we could hardly bear to stand upon it." He sums up his impressions: "We saw its eruption however, and took notice of immense masses of rock which it hurled upwards in smoke, and some of which were at least as large as the hull of our long-boat."[80]

In volcanism Forster found a visual language that was easily transferrable to the revolutionary turmoil in *Views of the Lower Rhine*, because inside the revolutionary also, there was a sudden and "unstoppable fire that erupted, consuming everything that resisted it. The revolutions that bring about forcible pressure are intense, fast convulsions that occur from the ground up—in nature, as in man. It is impossible at this point in time to escape such change." Until the late eighteenth century, volcanoes epitomized the unpredictable force of nature. James Hutton's *Theory of Earth*, published in two volumes in Edinburgh in 1795, was the first to assign volcanism a rationalized place within the whole of nature.[81] Forster uses the unsettling natural images of volcanism, which was not yet dispelled by theory at that time, to introduce the reader to eruptive political revolutions. Revolution is an underground happening of nature.

In nature, opposing forces wrestle with each other. They generate a spectacle, "beautiful and sublime in their most destructive effect. In the eruption of Vesuvius, in the thunderstorm, we marvel at the divine independence of nature. There is nothing we can do about thunderstorm-matter gathering in the atmosphere until the full funnel clouds threaten to destroy the earth; nor the volatile gases developing in the bowels of the mountains, which blaze the way forward for molten lava. We see it everywhere." The "primal forces of the universe" are entangled in a battle that must not end even "if the universe should halt and freeze!"[82]

The history of humanity does not behave any differently. Forster's inclusion of stark natural metaphors, such as volcanism, stands in historical opposition to the progressive idea of a gradual, continuous movement toward perfection. His claim is that the "century, like the human race, does not move forward at a regular pace, but rather moves in constant rotation."[83] By "rotation," Forster understands nothing less than "revolution." The conflicting forces at work here prove to be both destructive and productive. We have to withstand their struggle, since "just as growth ends in destruction, the development of one ability means the suppression of another."[84] Only when we accept the violence of transition as being necessary do we attain a new level of development: "All origins are chaotic, and the combating elements of chaos infuse revulsion or terror. When, however, the new creation emerges in quiet glory, then, we shall no longer remember the darkness and its storms."[85]

Forster regards himself as a worldly-wise naturalist, an identity that he sees as an advantage in understanding political eruptions. As an observer of nature and as a sociopolitical thinker on his way to becoming a revolutionary, he attempts to initiate the reader into what is to come. This is the bitter insight for which *Views of the Lower Rhine* seeks to lay the groundwork: without violence, nothing is going to happen.

Nature as Fate

The idea that the world is not static but has its own history is among the most exciting in modern science. When considered from both a natural and a cultural historical perspective, it was extraordinary. Before Alfred Russel Wallace and Charles Darwin, the theory of the constancy of species remained undisputed, enshrining species in a standstill of nature—with the exception of those species proven by fossils to be extinct. Yet, geological insight into the history of the earth's surface paved the way for the idea that nature itself may have undergone development, something that would no longer be in keeping with the biblical teaching of a single act of creation within the modest span of a few millennia. On the other hand, direct encounters with exotic peoples around the globe caused the idea to mature that even human cultures undergo a history.

Forster struck revolutionary sparks from this premise too. He sees humanity as a product of nature. In the easily overlooked text "Die Nordwestküste von Amerika, und der dortige Pelzhandel" (The northwest coast of America and the fur trade there), of 1791, he speaks of the "history of reason"[86] and traces the slow "progress of man from nearly vegetating to mere animal, and finally, to a rational life." Just the gradual "course of nature" yielded humanity's "full use of reason." In "Leitfaden zu einer künftigen Geschichte der Menschheit" (Guide to a future history of humanity), which Forster worked on in 1789, he describes nature as the force that maps out the "different levels of development" for peoples[87]—development is due not to the power of reason, which constitutes culture, but to the natural process that first yielded reason.

These are not purely anthropological or natural-historical considerations. Since 1786, Forster had been planning to write a handbook on natural history; now, in his "Versuch einer Naturgeschichte des Menschen" (Attempt at a natural history of man), he questions the natural preconditions for political systems, thereby allowing the leap from natural to cultural history. Political structures can in no way be attributed to the ideas of pure reason, but rather to various environmental influences. He charges "countless linkages of preexisting circumstances and events" to political constitutions: "How is not everything that happens now prepared from afar, such that we are unable to imagine the most insignificant link in the sequence without, at the same time, having to think of and expect a genuinely manifest but wholly different success?"[88]

It is worth recalling the basic premises of a physical anthropology, since they facilitate more serious reception of Forster's surprising remarks about political revolutions than perhaps initially suggested. What are we meant to understand when Forster writes of a revolution having come about "on its own"? This formulation crops up again and again: despotic rulers do not understand the processes of revolution, and so, "to their astonishment, the tremendous reversal of things finally gives way to change on its own"— "per se."[89] No trace of liberty, no mention of reason.

The key term, which Forster makes repeated use of in this context, sounds antiquated: he talks of "fate." The term seems to derive from the classical concept but can be quickly glossed over, as one is not inclined to take Forster at his word here. Who still speaks of fate in the age of the Enlightenment, which is to say, the age of self-determination, the critique of religion, and the resplendence of reason? He cannot mean it seriously. And yet Forster insists on it. His theory of "natural revolution" is thrown into sharp relief against the conceptual backdrop of fate.

In his view, there is a natural fate that still rules over our freedom. Because fate is so "conditioned to the vexing liberty of man, with which it imitates us," we become marionettes of nature's power: "Where we reckon to be free, it clenches us in an iron-clad need for convening circumstances, entirely independent of our whims, giving our lives direction, about which we can do nothing, and making us about as independent as the king of chess." Everything is connected to everything. We are spun into the net of reality and tossed into the torrent of history—mere marionettes in the course of time. Because in the "course of events"—a formulation that seems to anticipate Samuel Beckett—"everything that happens is determined by what has preceded it and by its relationship to the future." Forster's theory of "fate," or even "providence," as Oliver Hochadel puts it, is "no irrational superstition spiriting the world away,"[90] but rather a rigorous questioning of the Enlightenment's concept of subjective autonomy.

Fate is a motif that runs through all of Forster's writing about revolution. Revolution is a fate to which we must submit ourselves. "Nothing is freely done. We have seen where we cannot change anything until the circumstances come to pass that make our involvement possible. In this way, we are always in the palm of fate's hand; because the chain of cause and effect, which depends on the course of action and free choice by *others*, binds *us* as all-powerfully as the most fatal necessity." Forster's fatalistic trust in the natural course of things explains his revolutionary alacrity for radicalism.

It may have remained dormant as political events drifted languidly toward it, but Forster had already prophesied in April 1791: "So the envelope must finally come, however late but all the more total for it." Years before the French Revolution, it was Forster's conviction "that the fate of the people [had] already been meted out."[91]

In *A Voyage round the World*, the young Forster wrote of "the paternal love and unerring wisdom, which, in the plan of this world, has provided for the good of mankind." But he was already speaking to human intellect, which "unfortunately is too shortsighted to discover the true intentions of the wise creator everywhere in the works of nature." Over time, Forster mentions God less and less; nature supersedes. Forster remains loyal to the view that people are "only an instrument in the plan of creation!"[92] We are "inseparably linked," a conclusion that "would see everything in terms of cause and effect, and deny the possibility that any speck could have behaved differently from the way it did in fact move."[93] Forster speaks of "bitter necessity" and the "unavoidable laws of nature,"[94] and he laughs about the supposed wisdom of the politician, who "ultimately, only acts after the fact; consequently, he is not free, but dependent on a higher chain of things, which we can only stand by and watch. If the French Revolution shall have important results, they will not be hindered by these poor politicians."[95] "Nature" and "fate" become interchangeable terms for Forster. Fate is natural, and nature is our fate.[96]

Forster considered dependence on an "*unstoppable* fate" to be less a metaphysical speculation—he was not enough of a philosopher for that—than the presumed keystone of his theory of physical anthropology, which was only partially formulated. He saw humanity as being yoked to a web of related influences, which were not random but targeted: "Organization, upbringing, local conditions (in order not to use the word climate)—how much do they do? Not for our way of thinking and conceiving, but for our efficacy—left, right, straight ahead, upwards or downwards? God! And so it goes with the whole many-wheeled machine of the world, just like that and no different than how it is run."[97] Opposition is futile, "because everything is determined in advance."[98]

"Necessity" instead of "freedom"? Forster does not dissolve the tension between the freedom to act and determination by natural circumstances. "Beyond the limits of human understanding is freedom joined with necessity." In the "melee of the world," the "highest rank in the chain of events" exceeds our understanding.[99] Yet it is incumbent upon humanity "to bring *morality*" to the world's inestimable diversity, "while we work and suffer with *awareness*."[100] Neither the inner necessity of nature nor the possibility of our freedom can be understood. The relationship between providence and self-determination is difficult to determine and remains in abeyance. To Forster, freedom is not a form of autonomy that is independent of nature and grounded in reason, but by all accounts it is an outward form of nature itself, embedded in its entirety. To distinguish between nature and God, freedom and necessity, is to be accused by Forster of the "art of dissection"—an endeavor he gladly leaves to "speculative minds" and "trained metaphysicists," among whom he does not count himself. Forster emphasizes that "in the real world things always go amiss more than in our theories."[101] Accordingly, he measures the worth of his own reflections by their practical usability.

Thus, when Forster claims that revolution cannot be avoided, he does not mean it rhetorically in the sense of political inevitability, but quite literally: "In the greater course of human events, far more is involuntary than the proud, sentient animal wants to concede in his dreams of freedom. The revolution is truly to be viewed as a work of nature's justice." That sounds broad but is meant concretely. "The National Assembly," Forster continues, with Paris in mind, "did not think about going as far as it has; but the ironclad necessity of time and of circumstances has forced it."[102]

Our freedom stands ever at the service of a purpose determined by nature. "Nature or fate"—Forster names them in the same breath—specifies what is to be done; we merely have to be "the instrument carrying out the deed." In doing so, we are denied a view of the natural course of the whole. The "aspect of justice is too high for mortals here. What happens *must* happen."[103]

Here we have the glowing nucleus of Forster's revolutionary thinking. It cannot be made more explicable through theoretical charges. The point is precisely that this innermost driving force of nature can only be experienced, not adequately conceived. No plumb line of reason can suffice to gauge such depths. The hidden powers of nature are more primal than human reason.

Surely, Forster's faith in fate sounds like what might be called a "teleological philosophy of history," but any attempt to ennoble or defame Forster through such rubrication would be inadvisable. His thinking always remains a little too nebulous to be defined with any final clarity. In this, Forster remains loyal to his basic insight that nature is not to be seen through. It is not a puzzle that we solve, but rather a force that we should entrust ourselves to.

The driving forces of nature function like waterwheels in the underground of history: only with difficulty can they be brought to a halt. Few can surmise it, but there are signs of revolutionary change.

The Principle of Political Change: Fermentation

The events of the French Revolution were confusing and surprising to the contemporary observer. They would be impossible to sum up in a single word, but perhaps they can be dressed in an image. Images lend precision to the intangible, especially when exact understanding is not possible. Every intellectual of the late eighteenth century had in mind one event, which had shaken up the European world: the great earthquake in Lisbon of 1755. The earthquake not only destroyed the Portuguese metropole but toppled the idea of a world that is whole and arranged with care by God. This telluric catastrophe was suitable as a political metaphor for the collapse of the *Ancien Régime* and came to be applied to all subsequent revolutions. Goethe, who was sensitive to such shocks, saw the metaphorical transference bear out in his later impressions of the French Revolution, as well as those of his contemporaries: "As almost in an instant the earthquake of Lisbon caused its influences to be felt in the remotest lakes and springs, so we also have been shaken directly by that western explosion."[104]

The metaphor of the political earthquake evokes the abrupt upheaval, the sudden caesura, the shake-up of all familiar foundations. Disinclined toward anything volatile in nature or in politics, Goethe dubbed the July 1830 Paris revolution the "Paris Earthquake." Ascribing visual language to history gained currency in the context of various worldviews.[105] Even Karl Marx spoke of the "June Earthquake" in Paris of 1848.[106]

On the basis of his own immediate experiences, Forster displays a preference for volcanism over earthquakes—both causing tremors—as a figure of speech for revolutionary events. Yet Forster had at his disposal a deeper understanding of political change, which might, at first glance, be mistaken for another metaphor. In his books, essays, journals, and letters after 1789, a leitmotif of "fermentation" can be found throughout his political vocabulary. In "Historisches Jahrgemälde von 1790" (Historical portrait of the year 1790), Forster describes how, within just a few years, "in several European States, a curious fermentation has transpired, which was socialized by the attempt to give a new form to the constitution or bring it back to its earlier form." In other writings, too, Forster speaks of a "fermentation among the people," describing the "domestic fermentation in France" or accounting for the "general fermentation" throughout Germany. Fermentation is "inevitable with the introduction of anything new."[107] Forster observes that "fermentation occurs in learned minds, and that this fermentation propagates, too, to such an extent" that there can be said to be such a thing as a "spirit of this fermentation."[108] This spirit of liberty is "the salt, the seasonings, or the fermentation material" in a "newborn free state."[109] He speaks of the "fermenting agent of republicanism" and emphasizes that is it "impossible to remain neutral in a fermenting state." He considers "unrest, fermentation, revolutions of constitution" to be synonymous processes. For Forster, it is about understanding "which eccentric movements, which fermentations, in short, which *revolutions*," are driving history.[110]

This usage was likely proposed by Herder, who, in opposition to the rationalism of his time, set into motion an entire arsenal of organic metaphors and figures of speech for growth and metamorphoses, such as "seeds," "age," and "fermentation."[111] Forster limits himself, though, to the one term that is most telling of his understanding of revolution.

Today fermentation is used for the transformation that a dough mixture or a drink undergoes in a warm environment with the addition of yeast, milk, or acetic acid bacteria. Zedler's *Grosses vollständiges Universal Lexicon aller Wissenschafften und Künste* indicates what was understood by *Gährung* (fermentation) in the eighteenth century. It involves a development "through which the smallest particles of each body are moved in such a way that they are completely dissolved into themselves, made thin and mature, transformed, as it were, into an immaterial nature and being for the preservation and spoilage of the body." Although this description must sound vague today, it is precisely that blurring which makes it readily applicable. In a free state, according to Forster, the "fanatical sects, by being given time for fermentation, can finally transform into equable, wise, valuable citizens."[112] Through a political transformation process, fringe sectarian groups, which is to say, the smallest parts of society, can become useful to the welfare or woes of the body politic. It is that aspect of metamorphosis that is important here.

Forster has two possible developments in mind, which Zedler's *Universal Lexicon* cite: fermentation occurs "either on its own, or through the activation and help of a ferment," which is to say, a suitable ingredient. When fermentation occurs in people, it involves a self-activating development, according to Forster, which is as natural as the change observed in organic matter. When something decomposes, it comes into fermentation on its own, transforming itself, becoming inedible or refined. The idea of liberty can operate this way in a people, entailing an "overall spiritual fermentation in the masses of the human race." Forster anticipates self-activating changes in the political body of society, what he calls the "spontaneous combustion of reason in an entire people." Seamlessly, he confers scientific premises on the process of revolution: if "heterogeneous matter can ignite even by itself under certain conditions,"[113] then the tension between the different social estates can lead to upheaval. For the genesis of a revolution, it is "often sufficient for one to think of it as something easy," he quotes approvingly from the *Mémoires de Monsieur le Cardinal de Retz*. Forster equates fermentation with "dissolution,"[114] as a process that runs itself.

On the other hand, people can be so lethargic, so apathetic, so entrenched, that the revolutionary may have to be the agent that sparks fermentation. Because wherever a society persists in a state of bondage, where "grossly sensual pleasure and mechanically learned truths, formed not by one's own thinking or experiences, are the mainsprings of all activity—an external stimulus is needed in order to incite apathetic minds to resistance, and to give them a previously opposing direction, in which they can now persist just as blindly as before, and with a dull obstinacy, easily mistaken for energy at first glance, and the clever leader, who mixes into the masses as the leavening agent that brings them to fermentation, serves as the instrument."[115]

Both approaches share a common premise: fermentation processes of this kind are natural, organic—and, once they have begun, unstoppable. Forster sees nature as a dynamic whole and understands it to always be in flux, a play of forces. Embedded in these dynamics is also humankind. "To think that we are not for the sake of peace and warmth, but that our endowments and powers must develop themselves, and they develop best when everything is not so precisely weighed out, so perfectly balanced; they enable efficacy through pressure and counter-pressure, through compulsion and necessity, through compassion and fermentation." As was already apparent in *Views of the Lower Rhine*, Forster is less surprised than his contemporaries by revolutionary developments. For him, the artifice of hierarchy found in absolutist monarchies is in opposition to the flux of nature. A "political mold" will break down the despots.[116]

All of this could be taken for mere metaphor, for bold language invoked to suggest an understanding of diffuse political process. Forster suspected it might be so understood and rejected that understanding in a footnote to

"Parisische Umrisse" (Parisian sketches). A revolution has "all the signs of a severe illness, through which nature rids the body of strange or foul matter, which, in too great a quantity, causes first general solidification, and then a dissolution that is just as general." That general dissolution is precisely what Forster means by his broad reference to "fermentation" as a result of "political mold." "This is indeed more than a comparison," Forster emphasizes; "it is similarity, kinship, agreement between material and moral nature."[117] This cannot be underestimated: Forster assumes "agreement," a direct reference from the material to the moral-political. Revolution is a process powered by nature. With this premise, Forster remains true to his physical anthropological approach in his guiding vision of the political. To deem it mere metaphor would amount to misunderstanding.

However, Forster's guiding concept of fermentation possesses a metaphorical excess. His organically oriented vocabulary for the political is precisely chosen, yet it expresses an opposition to the mechanistic view of the state. This dualism between organism and mechanism is groundbreaking for its departure from political conceptions at the time of the Enlightenment.[118] Organic metaphors for the political are not foreign to this tradition: Aristotle compared the state to a human body, as did the Catholic church in recognizing Christ as its head. Although the metaphor of the world as a machine was already virulent in antiquity, Hans Blumenberg writes, it was the metaphor of the modern clock mechanism that would lend expressive power to "the nondescript and unspecific expression *machina mundi*."[119] Thomas Hobbes was the first to apply the machine metaphor to the political realm, writing in *On the Citizen*:

> As in an automatic Clock or other fairly complex device, one cannot get to know the function of each part and wheel unless one takes it apart, and examines separately the material, shape and motion of the parts, so in investigating the right of a commonwealth and the duties of its citizens, there is a need, not indeed to take the commonwealth apart, but to view it as taken apart, i.e., to understand correctly what human nature is like, and in what features it is suitable and in what unsuitable to construct a commonwealth, and how men who want to grow together must be connected.[120]

The individual becomes a cog in the state machine, which is rationally designed and can be operated by its highest ruler. Even Forster emphasizes that a machine is something that is "set in motion by another," while the moral, thinking, free man determines "himself through rational arguments."[121] He sees the danger in "a mechanism that deadens hearts and senses"[122] and gains the upper hand by driving back nature.

Insofar as the machinery of the state epitomizes the rationalization of the political, Forster defends the impenetrability of revolutionary potential. In mechanistic thought, the revolution has no place, yet a mechanism runs with complete uniformity like a clock once it is put into gear. Revolutionaries, though, are not cool strategists of the state. Rather, it is assumed "that nature determines such rare people for herculean work, for mighty effects, for great deeds; that nature must have fulfilled that consuming blaze of passion, which, once the epoch has withstood its first fermentation, will be led by indwelling spirits, defeating insurmountable difficulties and attracting the admiration of their contemporaries and posterity alike. Mistakes, aberrations, even crimes, are conceivable here; whereas vice remains fully barred, provided that the moral powers of such characters strengthen themselves through sensation's unlimited power to assimilate, and provided that they mature into the dominion that is their due."[123]

Forster's concept of fermentation as the principle of political change connects to his central message in *Views of the Lower Rhine*: Wherever there is fermenting, bubbling, festering, activating, and transforming, where, in other words, organic forces are at work, proving themselves to be more efficacious than reason, the moral powers of the revolution can remain intact, as long as they are guided by feeling; nevertheless "mistakes, aberrations, and even crimes" are to be expected. It is natural for there to be violence in revolution—even if it is a crime. Reason cannot hamper the forces of nature. Forster sees revolution as the natural flip side of reason.

French Liberty in Mainz

When Forster took up a librarian position in Mainz in April 1788, there was no fermentation to speak of in the electorate. Surely, there were a handful of professors who saw themselves as being attached to the critical spirit of the Enlightenment. Mainz was not exactly a backwater. However, its political structures could not be distinguished from those of the rest of courtly society. Germany was not a united nation, but rather a patchwork of small states, denominationally and culturally different, without a central unifying power, even though the Holy Roman Empire of the German nation persisted until 1806.

Forster was surprised to find how rigidly courtly etiquette was adhered to in Mainz. "You wear a tailcoat to the Elector's even if you have already been introduced to him." The estates were still neatly separated, hierarchy among the nobility remaining unshaken. After arriving in Mainz, Forster introduced his wife to the familial isolation to be expected, given their simple living conditions: "For even if we were invited to high society, they will not be calling on us in return."[124] His modest income as a librarian did not permit him promotion to the higher circles. "Now and then a foreigner, and a few friends, that is my society and my refreshment." Life plays out in private, "because," Forster writes to Schiller, "no one outside of our circle understands us."[125]

In spite of how enlightened the realm of ideas had become, Forster still had to move through a world that was conditioned to authority. The disparity between the elector and his attendants can be illustrated by one small detail. In a letter to the elector concerning library facilities, dated September 9, 1792, Forster requests "that Your Electoral Highness might deign to most graciously decide to grant permission to the Electoral University to erect in the library halls a portrait of *His Electoral Highness* in marble, on a lovely pedestal, with a grateful inscription to the University's second founder, its restorer, the father of its fatherland and the guardian and benefactor of arts and sciences, out of boundless benefaction for whose eternal memory the University is lavished and endowed."[126]

Goethe, by comparison, arranged for the putative skull of Schiller, who died in 1805, to be brought to the Anna-Amalia Library in Weimar, where it was unveiled in an official ceremony on September 17, 1826, and incorporated into Johann Heinrich von Dannecker's sculpture of Schiller's bust.[127] Although this had to be arranged with the Grand Duke Carl August— Goethe secured his "order and authorization" in advance—not much time was lost in the process. The difference between these comparable scenes serves to illustrate the changing times: in the library of prerevolutionary Mainz, Forster was still seeking out for tactical reasons a symbolic enthronement of the elector as an institutional representation of the intellectual world, whereas Goethe's sovereignty is attested by the ease with which he brought a relic of the peerless German representative of liberal thinking to a site of knowledge and the Enlightenment. While Goethe requests the "continuation of favor and good will" in his letter to the grand duke, Forster ends his petition with the words: "I herewith remain with deepest awe Your Electoral Highness's most subservient university librarian, Georg Forster."[128] The difference comes down to one point: Forster was still an underling whose political will was dwindling in the eye of the elector's persisting power, whereas Goethe was operating on equal footing, even though he was inclined to observe the rules of the game of decorum.

The Weimar poet might have engaged lightheartedly in this game of representation, but it would have been harder for Forster. His father had cheekily responded to the chamberlain's question whether being granted an audience with King Friedrich the Great at Sanssouci Palace had left an impression on him: he had already met five wild kings, and in Europe, two fully domesticated ones. This anecdote calls to mind the advantage of immediate experience, as possessed by the two world travelers: they had met kings from the other side of the world, come into contact with an array of societies and moral codes, breathed the air of the whole wide world—making the stuffy formalities of European courts seemed unbearable. "Everything here is empty and flat, and askew, too," Forster says of the conditions in Mainz.[129]

The journey along the Rhine with Humboldt brought only temporary change. Financially, it was unsuccessful. Forster had hoped to find a suitable publisher for the botanical descriptions he had written during his voyage around the world. This effort turned out to be a "complete failure." As an "ever humble servant," Forster turned to King Friedrich Wilhelm II of Prussia in February 1792. He had been the only one from whom Forster had been able to "beseech mild support" for his botanical descriptions[130]—alas, without success.

After he returned to his day-to-day work in Mainz, it was only a matter of time before the old displeasure set in again. Forster writes in 1791, "It is as if everything I touch turns to water—nothing flourishes. The more I work, the more I hope to gain, the more the anger melts in my hands. And now I stand here with empty hands, unable, as before, to work, and yet in no position to help my household manage if I do not continue in my previous toils." He is extremely productive: besides writing *Views of the Lower Rhine*, Forster translates a Sanskrit play, *Sakontala or the Fatal Ring*, having come across the English translation during his journey with Humboldt; he writes important essays such as "Über historische Glaubwürdigkeit" (On historical credibility) and "Über den gelehrten Zunftzwang" (On the obligation to join a scholarly guild); he completes twenty reviews in this one year alone, and he works on the laborious "Geschichte der Englischen Litteratur" (History of English literature), to name just a few examples. Yet, he complains: "I am sapped of strength, my body is not capable of any more exertion, my spirit grows weary."[131] It would not have taken much for Forster to become part of the local color, trapped between dusty bookshelves and barely feasible commitments to his publisher. He "took on just too much, and did not account for illness and for the time that certain reviews cost him." His life and thought were at risk of becoming lost in the daily, especially since there were no signs of that stimulating unrest that he had observed elsewhere during his Rhine journey. "I bathe in the Rhine while my life drones on," he writes to Jacobi in August 1791.[132]

A few months later, everything was different. Since the Great Revolution in France had begun on July 14, 1790, domestic and foreign political tensions had been mounting. Prussian and Austrian counterrevolutionaries were still hesitant to defend the old power relations by sending in troops against France, while some Jacobin representatives from the National Assembly in Paris—not to mention King Ludwig XVI—were already arguing for war at the end of 1791. Given the poor state of the French troops, the king might have been hoping that a French defeat would also mean the end to revolutionary activities and would permanently rescue his monarchy. On April 20, 1792, he declared war on behalf of Austria and allied Prussia.

With the change in French foreign policy since the end of 1791, the idea of an *expansion révolutionaire* was linked to the revolutionaries, as Jacobin proponents of war sought the protection and proliferation of republicanism. Volunteers serving under the French soldiers became known as *apôtres de la liberté*—defenders, if not apostles, of freedom. They were mobilized to the border regions, and soon the French were at the Rhine. "Everywhere between Mainz and Koblenz is teeming with the French," Forster reported in April 1792. "The entire Rhine region is bunged full of them; all the pubs are filled, making any gaiety impossible for the people of Mainz. That would be tolerable, but they cause the prices to increase for all of us; everything costs twice as much as usual."[133] The consequences of French expansionist politics were becoming increasingly apparent. "Our local situation is starting to turn critical," Forster wrote on April 17. A few days later, he put down his pen, as it were, and noticed that the spirit of resistance against the French was beginning to grow in Mainz, too: "There is not a literary thought to be had here. Soon they will likely be making the sounds of swords from our ploughshares; because from one day to the next, only a clear and defiant language is spoken against France."[134] By the end of the month, he announced, "The war has now broken out." Mainz was safe, though. There was nothing to fear, as the war would not play out in this area, but Forster, attuned to what was coming, wrote, "Things that cannot be changed, one must accept, like storms, frost and snow, rain and stormy weather."[135] On October 21 French troops, under the direction of General Adam-Philippe de Custine, seized Mainz.

There was hardly any fighting around the city, just smaller firefights and a brief negligible bombardment. Given the superiority of the French troops, the Mainz War Council decided to unconditionally capitulate on October 20.[136] Many Mainz residents had already left the city days before, as did the elector, "who departed in the still of night at nine-thirty and had his wagon stripped of the coat of arms." The tide was turning. "We have been in French hands since yesterday," Forster wrote to his new publisher, Christian Friedrich Voss, "since six o'clock in the morning," he specifically noted. That same day he followed up with another letter, in which he sought to assess what the French occupation meant: "Mainz will now become an important political center; because from here on out, French operations will make toward Germany." Forster found himself unexpectedly at the center of political events. The victim, just days before, of a life rippling with boredom, he now regarded himself as witness to one of the "decisive epochs of the world." He called upon his fellow citizens: "Germany which traces our every step, and the world which will judge us, demand action."[137]

Although Forster had been merely an observer of revolutionary upheaval, the opportunity now presented itself for him to have a hand in history. Just months before, he had shown restraint: it did not occur to him "to predict an *upheaval* that I myself did not desire, and what's more, believed would be a great misfortune for Germany. As a result, I mustered all I could to ward it off." Although he admits he is "rather for than against the Jacobins, however much people may rave against them," he is convinced that Germany is "nowhere near ripe for change in its constitution."[138] But Mainz is now occupied by the French. By the beginning of November, they had captured the region between Landau and Bingen on the left bank of the Rhine; on the right bank, the occupied region was marked by an imaginary line stretching from Frankfurt am Main, Friedberg, Weilburg, Limburg, Nastätten, to Lorch.[139] This opened up a world-historical moment for Forster that was being acted out on a small scale, and for which the rest of Germany did not yet seem prepared. Forster could now openly profess that he "was never en enemy of liberty" and that it was impossible "to fight against freedom." The times were changing, and Forster rediscovers among the French troops the equality that he had first found so intoxicating in Tahiti: "officers and commoners like brothers, one heart and one soul, eating together at one table in the pubs." The hollow ceremonies of the ruling powers vanish into nothing. The nobility cling fearfully to their dwindling privileges, as Forster acerbically remarks: "Cowardice and nobility have begun to become synonymous." It is time to act. Forster recognizes that "the crisis is here and one must take a side."[140] He becomes a revolutionary.

Franz Dumont, an expert on the Mainz Republic, has apodictically declared, in a critical departure from Marxist readings, that "Forster's significance to the Mainz Republic is generally overestimated."[141] That may be true; Forster was by no means responsible for the introduction of republicanism right from the start. On October 23, 1792, just two days after the occupation of Mainz, the Jacobin Club Society of Friends of Liberty and Equality was founded under the motto "live free or die" in the *Akademiesaal* of the electoral palace. Forster hesitated; he did not join the club until November 5. That may not seem very long, but politically charged times are counted in days, if not hours. Forster engaged only gradually in the upheaval.

Dumont's assessment misses the mark, however, in its assumption "that Forster saw to some extent a moral process in the revolution, which had to be expedited by individual and societal 'Enlightenment.'"[142] It is true that Forster did not want to relinquish the connection between the Revolution and the Enlightenment, because it was "the surest way to enlighten people about their actual situation and rights; the rest [would] come about on its own." But this merely scratches the surface, as it were. If the Revolution is to be thought of as natural, as an event fed by the sources underlying reason, the Enlightened revolutionary can only place himself in its service, no more. To put it another way: a revolution is not caused; it comes about "on its own." Thus, Forster initially persists in his position as an observer until the time comes to swim in the current of history. "In a few weeks, Mainz and its surrounding area will surrender completely to the French constitution," he writes in November 1792, "and the strangest thing of all is that it will occur without an upsurge, without enthusiasm, taking place as completely as if it were entirely something taking place that happened on its own and had to happen. There is nothing I can do but go along with it; this calls for cleverness."[143] From the moment he decides to go along, however, he becomes a prominent figure of the first republic on German soil, because he is sure of an audience beyond the borders of Mainz in a way that none of the other Jacobins are.

There may also have been personal reasons for his hesitation. He must have regarded his marriage as failed, owing to his wife's affair with Ludwig Ferdinand Huber. "I reckon," he admits to his friend Lichtenberg, "I've aged by at least twenty years—and not in the better sense of the word. I feel more dead than I should, like a plant that has been damaged by frost and can never recover again." Yet he tried to hold on to his family and was even open to a *ménage à trois*: "I do not know why all three of us should not hope to survive this great time," he wrote months later from Paris. He was poised, once again, before an immanent void. His position was more precarious "than it [had] ever been." He admitted, "The Elector is no longer in a position to pay me my wage."[144] As an author, he faced a threat of being cut off from his publisher, who was outside the occupied region. In a letter to his publisher, Voss, he weighed his prospects a few days after the capture of Mainz:

> What do you think can be done in a position such as this? To leave my house and furnishings, which is to say, all that I have in the world, and roam about with my wife and children until we can no longer support ourselves—or remain here and seek to uphold the university, adopt French citizenship in order to lead them on a sound, temperate path, so that when there is peace after reunification with the German Reich, if it should come to that, it will not be detrimental to have ventured what can be ventured on this career path—am I to be misjudged, decried, mistaken for the head demagogue, and so on, since everything now stands on the brink of being resolved in the most violent ways and being led by people everywhere who have nothing, not even honor, to lose? I see that I must choose the latter if I am to be guided by a glimmer of love for the welfare of all, by some feeling of dignity within myself, by concern for those close to me.[145]

Days later, he joined the Mainz Jacobin Club, and on November 15 he gave his first major speech, "On the relationship of the people of Mainz to the French."

Forster quickly found his tone: his speeches on revolution are solemn but simply given and are genuine in their central argument. Once again Forster presents himself as the opposite of a highbrow author and orator.

"What were we just three weeks ago?" Forster asks in his first major Club talk. "How could this wonderful metamorphosis have happened so fast, from the depressed, ill-treated, tacit subjugation of a priest, to raucous, free citizens rising up, to dashing friends of liberty and equality, prepared to live free or die! Fellow citizens! Brothers! The power that was able to transform us, can also meld the French and the citizens of Mainz into one people!" Forster's speech does not spare the familiar rhetoric of the revolution, when he invokes kindling the "holy flame" and warns against the "abyss" and the "pools of hell" that mark a return to the old systems. His speech is not lacking in mordant provocation, either, when he calls aristocrats "degenerate, feeble-minded, privileged."[146] Yet he does not fail to formulate clear aims. The commander-in-chief of the French troops, Adam-Philippe de Custine, had already recommended a republican constitution and confederation with France in a speech given in Mainz at the end of October.[147] Forster unreservedly takes up the cause, professing that "the freest constitution appears to be the best." For this, one must fight "with bravery to the death," and "individual persons cannot be taken into consideration."[148]

That was meant quite literally. In a letter to Huber, Forster writes that he now finds himself in "more complicated relationships." "[I see] before me the necessity to act as if I were the only person in the world who can be counted on. I can, and I want it." His wife and children were sent to Straßburg for safety. "I have decided upon a matter," he writes to Soemmerring, "for which I must offer up my personal tranquility, my studies, my domestic happiness, perhaps my health, my entire fortune, perhaps even my life."[149]

In these first days under French occupation, Forster experiences something new. As a writer, he has always worked through the indirect medium of paper and could only diffusely make out his audience. As a political orator, he now has the experience of immediate influence, which, of course, cannot be compared to academic lecturing. "It is astonishing how one can affect people," he says with surprise. He takes a liking to the idea of being able to directly influence people through words—no longer isolated readers, but crowds, because "these masses will indeed be moved, with the right lever."[150]

Except that the masses did not quite want to be moved. Forster himself seems to heave a sigh. "I am now an underling—no, that word is banned here—a citizen of the French Republic," he writes to his father, adding that he has "given speeches to applause" at the Jacobin Club. He even proudly informs his publisher Voss that he is now a "free citizen." This was the spirit of the time: a "deeply humbled people" will become accustomed to "lifting up their heads and feeling like human beings who have been set free."[151] Forster no longer wants to deny himself the righteous political path.

But there were already critical voices appraising the proclaimed new epoch. "Our liberty is near," Forster wrote to August Wilhelm Iffland, the famous actor who played Franz Moor in the premiere of Schiller's *The Robbers* in Mannheim. Iffland replied with reservations: "I confess that I do not look forward to this era, and I will be saddened when it comes to pass! If the current form is changed, the first estate shoved off its bench, and the third estate spreads in its place, what do we gain?" Iffland feels repelled by the churlishness of the third estate and sees a strain of despotism in the absent masterminds of revolutionary events: "Almighty! Save Germany from the German intellectuals!" Iffland was not alone in his reservations about the new Zeitgeist. Lichtenberg spoke of "liberty influenza." Forster was serious about it, though: "To be free is to be human,"[152] he writes in defense of the "poorest peasant," deeming such folks "sacred and worthy" against Iffland's objections.[153] Forster in no way conceives of himself as one of the intellectuals, who sought to realize pompous ideas, but rather as someone who placed himself in the service of the development of the human race, because only someone who was free could "cultivate his mind to perfection."[154]

He was all the more surprised when the November and December 1792 elections for a new constitution turned out to be a failure. Although Forster could boast that from Speyer to Bingen, "nearly all votes had been unanimously cast for the adoption of the French constitution and accession to France," this tally was based on approximately only 8 percent of locales that voted. In Mainz, "greater tepidness" could be detected, which could also be called "unwillingness for revolution," since the Jacobins were not able to win even one-tenth of the citizens' votes.[155]

Forster reacts tetchily. "German inaction and indifference are to be spat at," he writes to his wife. It does nothing, he complains to his publisher Voss, "to coax" the peasants out in the countryside, and Germany is not at all "ripe for any revolution." But a natural revolution is not a matter of free choice. Thus, Forster only doubles down on what today might be called "public relations," namely, in the *Neue Mainzer Zeitung*, which began publishing in January 1, 1793. But this idea had long since taken root in him, and it was ready to transfer the pressure of natural events to revolutionary outlets. If the Germans still do not see eye-to-eye with world-historical processes, they need not be taken into consideration. "In the end, we will surely have to take mercy on them while issuing the command that they should and must become free; that ought to suffice."[156]

The Mainz Republic

"Forster's conduct will certainly be criticized by everyone, and I can see in advance that he will derive shame and regret out of this business," Schiller wrote on December 21, 1792, exactly two months from the day the French occupation of Mainz began. And already Forster's reputation was at risk. Schiller's remark reveals him to be patronizing and indifferent to the events in Mainz. With world history and the Greek gods in mind, he is unable "to take an interest in the Mainzers, because all their actions do more to signal a ridiculous obsession than healthy principles, which is not compatible with their conduct against the dissidents."[157]

Schiller did not understand Forster. Unlike Goethe, he did not have an intuition about the inner motivations of his local revolutionary world traveler. Forster had long before abandoned "healthy principles," because, unlike Schiller, he was not a Kantian. For him, the revolution was a matter of natural feeling, of sensation, of the heart, perhaps of reason, too, in the end, but not primarily. "We live in a strange time," Forster writes in November 1792, "when people can hardly be judged anymore, if they are to be sized up only according to their external circumstances; when the standard by which we otherwise care to measure each other, I dare say, must be shattered, and only humanity remains left. Principle, character, change, career—one can go mad on all of them now; the heart, where it is present, seems to be the only thing remaining among honest men that is the true *point de ralliement*"[158]— the true rallying point. This is to say, Schiller had no clue about revolution. Revolution is a state of emergency, which cannot be tamed by the powers of reason alone, because "at every stage of civilization which the human race has achieved, or will achieve in the future, wants and passions are the mainsprings of all constructive and destructive actions alike."[159]

Forster is not unreceptive to flights of reason. During his tour with Alexander von Humboldt, he came upon a book he found inspiring: "I have obtained from England," he writes his publisher Voss, "an admirable piece of writing by the American Thomas Paine, the famous author of *Common Sense*. It is called *The Rights of Man* . . . four editions are already out of print." In this pamphlet, which was published in two parts in 1791 and 1792, Paine defended "*the unity of man*, . . . that men are all of *one degree*"; thus, "all men are born equal, and with equal natural rights." Forster might have heard in this a personal echo of his debate with Kant, but he saw no possibility for publishing such programmatic writing in Germany: "It is *so democratic*, however, *I* cannot translate it because of my situation." Its "daring republican language" was reputed to spark "convulsions" among liberal Britons at the mere mention of the author's name.[160]

To translate Paine's pamphlet, Forster enlisted Margarete Sophia Dorothea Forkel, a translator who had assisted him with his work in Mainz, but his publisher Voss declined. Disappointed, he informed his comrade-in-arms of the rejection, "shamefaced," "as if he had spoken a death sentence." It is a testament to Forkel's commitment that Voss was brought around—"When you see the book," she wrote him, "you will not be able to do anything other than print it, even if it should be high treason."[161] Forkel translated, and Forster wrote the foreword. Evenings they would go through the drafted pages together "most punctually," even jubilantly as though they were "trophies" and Forster were "carrying off the palm," his accomplice noted.[162]

Paine's writing obviously struck a nerve with Forster. It was not the idea of liberty and equality borne out of reason that affected him, but what Paine expressed about the rights of man and how that idea resonated with what he had already seen and experienced during his voyage around the world. "The lowest man in the nation," Forster had written about Tahitian society, "speaks as freely with his kings as with his equal, and has the pleasure of seeing him as often as he likes. . . . The king at times amuses himself with the occupations of his subjects, and not yet depraved by the false notions of an empty state, often paddles his own canoe, without thinking such an employment derogatory to his dignity." In the foreword to the German edition of *Rights of Man*, Forster enthuses about Paine's speaking with as much "candor about kings as he does about common people."[163] It is this consonance between Paine's intellectual freedom and Forster's own experiences that must have made him feel understood. Finally, so it must have seemed to him, what was already being lived on the other side of the world would exist in European thought, at least.

Forster describes a key scene that occurred at the center of the Revolution and reminded him of his experiences on Tahiti. Forster and Humboldt's tour of the Rhine ultimately brought them to Paris in July 1790. They did not have much time. "On our quick flight through France (we were only in Paris for three days), I reckon to have observed that the revolution, more

than Germany would believe, has been consolidated. Everything is newly organized already," Forster determines. In Paris they attend the preparations
for the first anniversary of the storming of the Bastille. Thousands worked
hand in hand on the Champ de Mars to erect an amphitheater. "The sight
of the people's enthusiasm, and so eminently on the Champ de Mars, where
preparations [were] being made for the large national celebration," inspired
Forster. "It is heartening, because it runs so universally through all classes
of people, and it affects so purely and simply the common good with disregard for private advantage." And then the unimaginable happened: Ludwig
XVI appeared "without bodyguard, without entourage, alone in the middle
of two hundred thousand people, his fellow citizens, no longer his subjects.
He took the shovel and filled a wheelbarrow with dirt, to loud cheers and
applause by the crowd."[164] The king as a man of the people—Forster had
already experienced this. As a revolutionary, his alacrity for radicalism may
have fed off the symbolic significance of repetition.

Egalitarian enthusiasm was out of the question in Mainz. No revolutionary mood had taken hold of either the city or the rural populace, despite
enlightened speeches and the symbolic planting of liberty trees. "All has
remained fruitless," Forster summed up in January 1793,[165] and the reasons
for it lay—in part, at least—in the unsuccessful change in political institutions.[166] Although Forster proclaimed "that the rulers belong to the entire
people,"[167] there was no proper representative body to give expression to the
will of the people. The disaster of the earlier vote on the constitution, which
at least in the rural areas had not been exhaustively conducted, confirmed
this. At the same time, the attempt to gather support from those who stood
to be liberated was proving to be a fundamental problem of the Revolution.
The old structures were criticized but were not radically abolished. How
were those who were accustomed to the old ways supposed to be able to
come around to this new freedom? How long was one supposed to wait for
the subjugated to approve their own liberation?

The impulse for change once again emanated from the French. General Custine was hard-pressed by the Prussian recapturing of Frankfurt on
December 2, 1792, much to the jubilation of the city's populace. Martial law
was declared in Mainz. Revolutionary liberation was at stake. The guiding
practice of a liberal occupation politics was now abandoned in favor of an
increasingly "authoritarian democracy."[168] The Paris National Convention,
unsettled by the recapture, issued a decree on December 15, 1792, that newly
modified foreign policy directives and essentially limited the right to self-

determination by those liberated from despotism.[169] The pressure for French freedom, which Forster had considered necessary but had not vocally demanded until now, began to mount.

To compel democratization, the occupied regions were "municipalized," or split up into administrative regions according to the French model. Citizens had to perform a "citizens' oath," as outlined in the election regulations, which Forster cosigned on February 10, 1793. Citizens also had "to vote on their own municipality, not in perpetuity, but rather for only two years."[170] This elected administrative body was supposed to, as a rule, consist of two municipalities (governing bodies for cities) and one mayor. All municipalities then had to elect deputies to the National Assembly.

Meanwhile, Forster had been elected president of the Jacobin Club and was crucially involved in all of these innovations. He worked incessantly and relentlessly. The population's willingness to follow municipalization left much to be desired, since it was understood as a coercive measure by the occupation—as was the oath. "The citizens of Mainz simply do not want to take an oath," Soemmerring wrote to Heyne, in a letter dated January 15, 1793. And writer Friedrich Christian Laukhard complained about how his fellow citizenry "are either talked or pressured into participating in a new constitution, particularly by *Georg Forster*. One must, whether one wants to or not, swear to the flag of liberty, plant liberty trees, and otherwise organize [himself] in accordance with the new system."[171]

The required citizens' oath made enemies out of those who were reluctant or had objections. Opposition to the revolution could be found not only beyond the city walls but from within the forming republic! As a demonstration of strength, gallows were erected in Mainz. Anyone who did not want to perform the oath would be expelled, or "exported," chased from the city gates. More than a few took this chance.

The pressure to succeed was high. It was not without theatrics that, on November 8, 1792, the Mainz Jacobins displayed two books in the Akadamiesaal of the Electoral Palace: one was a book bound in red Saffian leather embellished with a liberty cap and the French national colors; it contained a short text laying out the case for liberty and equality and the corresponding—French—constitution, but otherwise its pages were blank. The second book was wrapped in black paper and entwined in chains. This book, too, contained empty pages; on the first page was merely written "Slavery." Both books have traditions in their own rights: red books had been used since the late Middle Ages to document the town charters of free cities, self-imposed by their councils and citizens alike; whereas Freemasons had used black books to record the names of those who were excluded from their lodges. Now, all male citizens of Mainz who were at least twenty-one years old were being encouraged to register in one book or the other as a way of putting their convictions on record. The crude dichotomy of freedom and slavery, light and darkness, good and evil, friend and foe, was designed to force the fateful decision. Although the Mainz books have not been preserved, the number of citizens who registered was quite considerable: the red book boasted fourteen hundred entries within just a few weeks in 1792 — although only one-quarter of them were eligible to vote.[172]

Forster later became ashamed of this "intense pressure" by "Mainz despotism": "Is there really any choice when disgrace and abuse await those who do not go to the red book?" he asked. If someone had had the courage to register in the black book, that would have been a testament to "freedom and independence of the will." To the disgrace of the German aristocracy, not one of them had such temerity.[173] There are reports of four people having entered their names in the black book.[174]

In retrospect, Forster may seem to have been alarmed by the attempt "to probe the citizens' convictions."[175] During the days of the revolution, though, he was uncompromising. With regard to the Mainzers, he writes with an audible sigh, fanaticism and ignorance had "caused an obduracy among the residents, which is regrettable, but at the same time must be handled with the most unyielding severity." Even the villages in the countryside must "be handled with equal severity." He has no intention of relenting to the Germans' unwillingness: "Such a foolish, callous, mindless people they are!"[176] The peasants, as he himself said, were fond of him, and they let him "rather naively know" it.[177] Forster is not an intellectual who has cultivated a distance between himself and the simple folk, not even as a revolutionary. He is impatient, though.

On March 17, 1793, the republican hour finally struck. In the Mainz Deutschhaus—today the seat of the Rhineland-Palatinate state parliament—the Rhenish-German National Convention was constituted, "with open doors in the great hall," as Forster emphasized, so that, "in the presence of the free people," the deputies could endeavor to "compete for their business."[178] Forster was elected vice president, but he held a great deal of influence and was the ostensible head—"the lettered members of the Convention say I am the soul of it." On March 18, a republic independent from the German Reich was declared. The issued decree stated: "From this point forward, the territory extending from Landau to Bingen shall constitute a free, independent, indivisible state, obeying common laws founded on liberty and equality. The single lawful sovereign of this state, namely the free people, declares, through the voices of their representatives, all relations with the German Kaiser and Reich revoked." In addition to the twenty explicitly named, among them the elector of Mainz, all of the German imperial states, as well as their secular and spiritual bodies, were declared incompatible with the sovereignty of the people. "Their demands for this state and its parts are forfeited," said the decree. It continued, "All their rights to sovereignty from usurpation are hereby terminated for perpetuity."[179]

Forster had placed all of his prospects behind this one effort and did not rest. Politics is a strenuous business. He came "right back in full swing, . . . feverishly working," he wrote his wife. Friends and observers alike winced: "How it pains me to see Forster so ensnared, I cannot describe to you," Soemmerring told Forster's father-in-law Heyne. Witnesses could "scarcely describe in strong enough terms the degree of his vehemence." Said Soemmerring, "I believe he is not quite himself."[180] Soemmerring cautioned Forster to think of his safety, but it was questionable whether he would heed such advice. He was already aware of Forster's future—"for Germany, he cannot be saved."[181]

Forster may have sensed this, too. Thus, he sees the necessity of unequivocally seeking the accession of the Mainz Republic to the French nation. Now or never! On March 21 he gave a fulminating speech, "On the unification of the Rhenish-German free state with the French Republic," which was rife with apocalyptic notes. He proclaimed with reverberant pathos, "The terrible day of judgment has come; the final hour of tyranny has struck; the obduracy, blind rage, and impotent overexertion of their last ounce of strength are signs of their ghastly death throes. The dying beast now squirms at our feet." The "aggrieved humanity," which had already presented itself as an issue in Tahiti, "steps into full possession of its rights."[182] It was all so unexpected that the liberators would be clung to all the more fiercely. The free Germans and the free Franks were "henceforth an indivisible people!"[183] The Prussian-Austrian troops were already gathering outside of Mainz for the siege.

On the day of Forster's enthusiastic speech, March 21, the Rhenish-German National Convention moved to petition for unification with France. On March 25, 1793, ninety representatives signed the address to the National Assembly in Paris, petitioning for accession to the French Republic.[184] The unification request was to be presented on-site. On that same day, Forster, risking his good fortune, wrote his wife: "Today, dear Therese, I depart—with two other delegates, Potocki and Lux, to accompany Deputy Hauss-mann from the Paris National Convention—for Paris in order to deliver the request for unification and accession, as well as an address on behalf of our local convention. I shall likely be back in three weeks."[185] Forster never set foot on German soil again.

Experts on Subterranean Passages: Forster and Goethe

The fate of the Mainz Republic was swiftly sealed. Starting in April 1793, Mainz was besieged by Prussian-Austrian coalition troops. In June the bom-bardments began, and by mid-July capitulation negotiations were taking place. On July 23, the republican specter was over. The free state had existed for all of four months: for the historian accustomed to thinking in centuries, a blink of the eye. Like a shooting star, the idea of liberty and equality rapidly flew past and burned up. In Goethe we have a perfect eyewitness, one of un-rivaled perceptive faculty, to this moment of the republic's quashing.

The idea of Goethe on horseback at the front lines is still disconcerting to imagine. Between bomb explosions and exchanges of fire, among soldiers and the wounded, he balanced a view of both the horrified public at a mo-ment of misery and the history of his epoch. As the Prussian general, Duke Carl August of Saxony-Weimar-Eisenach supported the campaign against France in 1792, as well as the siege of Mainz one year later. Goethe accompa-nied him. He wanted to expose himself directly to the chaos of war. The poet as war observer! No one could have expected it. "Goethe in the army! What profanation!" Heyne was appalled. Not until three decades later did Goethe record his experiences in "Campaign in France 1792" and the "Siege of Mainz"—not as a factual reporter of history, but as the medium for immedi-ate historical experience.[186]

Because Forster was in Paris, he was not there to see the recapture of Mainz. You might even think that Goethe had taken Forster's place by de-sign, as though, for this one moment, the task of faithfully experiencing the course of history had been handed from Forster to Goethe—the Weimar poet reporting stirring events in Forster's stead. Beyond the superficially obvious differences, their shared notion of nature's hidden driving forces serves as a connection. They were united in their conviction that nature should be understood as a moment of the political. Forster found the most beautiful formula for Goethe: he was "consecrated in nature," and he himself was filled with love for "holy mother nature."[187] Both Forster and Goethe were mistrusting of the veneer of pure reason. They did not settle in the upper floors of civilization, but they knew the instability of the foundation on which it stands. This shared insight was more influential than any of their differences in opinion. The two enjoyed a mutual appreciation, which is surprising when you consider the rebel and the poet—one sensitive to rumblings, the other ostensibly opposed to the revolution, yet each one devoted to the other. From how sparsely their encounters have been documented, one can conclude that it was Goethe who appreciated Forster more than Forster was capable of fully appreciating the polymath Goethe. To speak of a friendship between the two would be going too far—one need only to consider Goethe's uninhibited enthusiasm for Alexander von Humboldt to detect the difficulty involved in his relationship with Forster. Here we have the clash of two antagonists who nevertheless thought of themselves as connected.[188] A brief chronicle of the times when they met reveals that their encounters were quite fraught.

Their first meeting in September 1779 barely left a mark. At least Goethe does not mention it. No remarks by him about Forster, not in his journal, not in the letters that followed, not even in his "Letters from Switzerland," which recount his second Swiss journey with Duke Carl August from September 12, 1779, to January 13, 1780.[189] Their first encounter was, you might say, inauspiciously timed: it occurred right at the start of Goethe's journey, during his stay in Kassel from September 14 to 16. Yet it drove Goethe to the Alps. Having traversed them once before, he knew of the imposing mountains that awaited him. He wanted to tackle the Furka Pass to Gotthard on foot, a formidable undertaking with the onset of winter.[190] The risk paid off. Goethe paid a subsequent visit to Johann Caspar Lavater, whom he revered at the time for his worldly wisdom. Meeting the world traveler Georg Forster was merely a prelude to all of this adventure and promise. They dined together with the duke, and on their last day, the three of them visited the *Antikenkabinett* as well as the art collection at the *Fridericianum*. That cannot have been as boring as it sounds, but the whiff of obligation could not be ignored. "I had to stay with them, eat with them, and right after lunch—which was served early—they took off," was how Forster recalled it.[191]

Even from Forster we get no immediate reaction to their first meeting. On the second of the three days that Goethe spent in Kassel, they did not see each other. Forster had more important things to do; he was "too busy."[192]

Thus, the impression remained fleeting. "I have seen Goethe, spoken with Goethe, but not enough to know him," he reported to Jacobi. Goethe questioned him about the "southerners," "whose simplicity he took delight in."[193] And an exchange about Jacobi's novel *Woldemar*, which Forster praised enthusiastically out of solidarity with his friend, turned out to be a faux pas, because Goethe did not share this enthusiasm. A few months earlier—unknown to Forster—Goethe had publicly ridiculed Jacobi's novel, nailing an example of the book to a tree. Goethe must not yet have been known in those days for his dusty classics and Olympian reticence.

Nevertheless, Forster's enthusiasm for Goethe sounds a little evasive and forced in its euphoria. "You know him," he writes to Jacobi. "You know what it feels like to only see him for a few hours, to only speak *alone* for a few minutes, and then, like a meteor, to lose him again. It goes unsaid." Not without pride, Forster informs his father of his acquaintanceship with Goethe, but he likens it to having seen a phantom: "Goethe is a shy, reasonable, sharp-witted man of few words; kind-hearted, simple in his being. Pah! Men who stand out from the masses cannot be described."[194]

Despite any tension caused by the delicate conversation about Jacobi's *Woldemar*, their first meeting appears in no way to have been marred. Goethe and the duke did not give any outright indication during their dinner together, and fortunately, Forster told Jacobi, he did not commit "any *sottise* [folly]." He confessed, though, "I spoke of great lords with great candor. I reckon there were a few moments when it took Goethe a bit of effort not to snort at my ingenuousness."[195] Impartially, Forster opined, "And undoubtedly they appreciated it."[196] Goethe ultimately overlooked their difference in literary opinion.

Three years later they met a second time—once again in Kassel, in early October 1783. Although they did not spend much time together, at least an amicable relationship seems to have been intimated. "Six weeks ago Goethe was here *at the court*," Forster writes Jacobi on November 13. "I saw very little of him, as we had different plans. He struck me as more serious, more reticent, more reserved; colder, more gaunt, and paler than before; and yet, with friendship and with a certain something that seemed to say, he was trying not to appear changed." Goethe, despite his reticence, went to some trouble for Forster. When Forster was planning a visit to Weimar the following year, the fact that Goethe "offered him a flat in his house" says quite a bit.[197]

They did not see each other again until September 1785. Forster and his wife stopped off in Weimar during their honeymoon. He had planned "merely to pass through Weimar" but now spent "two of the happiest days" of his life there. Goethe gave a "Greek dinner" one evening, at which Herder and Wieland were also present. Forster found Goethe "better than expected"; the poet was "good and lively" and "opened up his heart" to Forster, such that he felt "honored and grew fond" of him.[198]

It is important to note that this mutual regard grew cautiously and was only made possible by their final ghostlike meeting, which took place over two evenings in Mainz in 1792, on August 21 and 22. Mainz was not yet occupied by the French, but republicanism was in the air, at least in Forster's house. The counterrevolutionary armies, which Goethe accompanied as an observer, had already assembled. Goethe and Forster—and this had long been apparent—were standing on a different continent, world-historically speaking, where it was every man for himself. Yet they spent "two pleasant evenings" together with Huber and Soemmerring, as Goethe describes in *Campaign in France, 1792*: "There was no talk of politics, for we felt a mutual need to spare each other's feelings: while they did not altogether deny republican sentiments, I was for my part clearly making haste to join an army which was intended to put a definite end to such sentiments and their consequences."[199] Political opponents they were, at the same table.

Would Forster have actually held back his political views that evening? Hardly. In a preliminary sketch that Goethe worked on in June 1810, he was unsparing in his judgment of their meeting in Mainz. He refers to "great tension of republican sentiments" and writes, "I was uncomfortable in such company."[200] Precisely for this reason, however, the harmonious portrayal in *Campaign* provides important proof of Goethe's desire, beyond the very real tensions and differences, for friendship and solidarity with Forster.

The gulf between the two cannot in fact be overlooked: Goethe sees in Forster a "man with a decisive way of thinking," which in political matters cannot be good, and Goethe strikes Forster as being "quite aristocratic."[201] Goethe's comedy *Der Groß-Cophta* (The grand kofta), in which he brings his negative position on the Revolution to the stage, elicited pure outrage from Forster, when Goethe sent it to him in the mail: "I leapt when I tore open that seal and saw that it was the *Groß-Cophta*. And now! O what a falloff! This thing lacking salt, lacking a memorable idea, a developed sensibility, an interesting character; this flat aristocratic everyday dialogue, this vile scoundrel, this mere courtly rescue of the queen." Goethe struck Forster as a reactionary bumpkin; Forster could not recognize the signs of the times anymore. His devastating judgment of Goethe, as expressed to his father-in-law, was even more explicit. "[The play is,] alas, everything that we otherwise delight in about his work; no spark of spirit, imagination, aesthetic feeling; everything is as flat as the Empress of Russia's shaman. Is it possible this man has been able to survive like this? Pity the ink and the paper!" And to his publisher Voss, he confessed that Goethe's newest work "scared [him] to death": "God save me if this is the way it should go for me, too! Better at times to quit than to plummet from the mediocre heights where Goethe stands to the most vapid depths. Indeed, there is not a line in the entire play that I remember or care to repeat—no imagination, no dialogue, no interest of any kind!" Would Forster have been able to cover up these misgivings at their meeting? It is no coincidence that Goethe was the one to seek Forster out in Mainz. And he puts up with him for "two pleasant evenings."[202]

Goethe would have been shrewd enough to recognize in Forster a contrary aspect of his own self. This is easily overlooked, as Goethe repeatedly emphasized his principal rejection of the French Revolution. His political ideal involved gradual, organic growth, not an eruptive break with all tradition. The French Revolution was the "most terrible of all events," he wrote, not coincidentally, in *Morphologie*.[203]

Yet, for all his publicly professed rejection, Goethe also affirmed an insight that might have come from Forster's mouth. After reading Jean-Louis Giraud Soulavie's *Historical and Political Memoirs of the Reign of Louis XVI from His Marriage to His Death*, published in Paris in 1801, Goethe writes in a remarkable letter to Schiller regarding the French Revolution: "It gives one, upon the whole, an extensive view over the rivulets and streams which, in accordance with natural laws—coming down from the many heights and along the many valleys—dash against one another, and finally make some great river overflow its banks and occasion an inundation which proves the ruin of him who foresaw it, as well as of him who was unaware of the danger. In this tremendous series of experiences we see nothing but nature, and nothing of what we philosophers should so much like to call freedom."[204] Political events as natural phenomena—Goethe and Forster were closer on this point than they ever could have known.

It is not an isolated case that Goethe's thinking should align with Forster's. In his 1815 festival play *Des Epimenides Erwachen* (The awakening of Epimenides), Goethe describes Napoleon's downfall as an imposing natural phenomenon, because had the winter not provided an end to the conqueror's Russian field campaign? Goethe uses geological features to symbolize counter-Napoleon forces in his play: east for Russia; the straits—or belt—of Denmark for Sweden; and the ocean for England:

> Rolling from the east, oncoming, like avalanches
> The ball of snow, of ice rolls large and larger,
> It melts, tumbling past close and closer
> waters that flow over all:
> It streams westward, then south,
> The world looks destroyed—and feels better.
> From the ocean, from the belt, rescue comes to us;
> All this works in happy chain.[205]

Goethe's descriptions of nature are much like Forster's "mere metaphors." Forster would have agreed with the wonderful formulation "The world looks destroyed—and feels better." Still, the destruction is natural. And nature halted Europe's demonic conqueror, Napoleon, in his tracks.

Goethe and Forster come close in their understanding of a natural politics but diverge on the actual formulation. Even Forster compares political upheaval to an avalanche. A revolution is a "natural phenomenon that is too rare for us to know its peculiar laws." "Restricted and determined not by the rules of reason," it must "retain its free rein." It is like an avalanche that "crashes with accelerated speed, making gains by crashing en masse, crushing every obstacle in its path."[206]

Despite the guiding idea of organic growth, even Goethe's conception of a natural politics is cryptic. He, too, assumes hidden driving forces, which govern our fate beyond reason and liberty. In a letter to Johann Caspar Lavater, eight years before the French Revolution, Goethe admits as much. "Believe me," Goethe writes in, a remarkable statement that cannot be overestimated, "our moral and political world is mined by subterranean passages, cellars, and cesspools. No one thinks and feels how a great city, in its connectedness and relations to its occupants, used to be. Only to [him] who has done some reconnoitering about this does it become more conceivable, when the Earth shakes for the first time, smoke rises over there, and here strange voices are heard."[207] "Believe me, what happens below ground is as natural as what happens above ground, and he who does not banish spirits by the light of day and under an open sky, does not summon them in a vault at midnight."[208] Goethe and Forster prove to be experts on subterranean passageways for politics writ large. Forster takes into consideration, as Goethe does, even if contemptuously, the "filthy subterranean canals" of the revolution's events. "It is not true," he writes his wife about the chaos in Paris, "that the driving forces are hidden, because I see them; but one must be in a position to be able to discern them. Our countrymen do not know the nation; it is a phenomenon of nature."[209] On this point, Goethe and Forster agree. The two are diviners of world history.

Goethe takes into consideration the dark side of reason, too, the reverse side of the Enlightenment. Revolutions are like earthquakes that surprise only those who are not sensitive enough to grasp the subterranean dimension of the political. The inextricable bond between Goethe and Forster can be recognized in this conviction; yet Goethe does not become a sympathizer of the Revolution: "All violent transitions are revolting to my mind, for they are not conformable to nature." However, he can agree with the French Revolution. It is no mistake that Goethe turns to metaphors for revolution in the preface to his *Theory of Colours*, when he argues for the indivisible unity of light as a principle of nature. The task, he claims, of rejecting Newton's established theory of *Optics* is akin to "razing this Bastille."[210]

Goethe's account of the siege of Mainz is as purely a historical report as Forster's *Views of the Lower Rhine* is merely a travelogue. Goethe wants to experience the unleashed power of politics writ large as a contemporary witness to history. On June 19, 1793, the bombardment of Mainz began. "The night was absolutely clear and full of stars, the bombs seemed to vie with the heavenly lights, and there were moments when it was really impossible to distinguish between the two." Goethe always keeps both the war and nature in view. When the Austrian-French boats began shooting cannon balls across the river, it was "something [he] had never seen before. When the ball hit the water the first time it caused a great splash that went many feet into the air; the first column of water had not yet come down completely when a second spout was thrown up, strong like the first but not quite so high, and then followed a third, a fourth, ever declining in height, until finally the ball reached the vicinity of the boats, skimming the top of the water and posing only a certain incidental danger to them. I could not get enough of this spectacle."[211]

As a result of the bombardment, "the collapsed towers, roofs with holes in them, and smoking ruins" left a "melancholy sight." And yet, in the siege of Mainz, Goethe sees that rare case, "in which misfortune itself promised to yield material for the artist." The "misery of a national capital on fire" offers a sight that reminds Goethe of Vesuvius. "So long as there was space enough to remain at a safe distance, it was a grand, uplifting spectacle," he wrote when he visited the active volcano during his Italian journey.[212]

For weeks, Goethe withstood the "most horrible scenes of war," drifting "through these terrible days like over a bed of hot coals."[213] The suffering was hard to fathom. On June 24, 1793, fifteen hundred people left the city, fleeing through the Wiesbaden city gate, but the attacking Prussians refused them passage. Stuck between the two fronts, they were "cruelly turned back." Goethe reports of untold sorrow: "The despair of defenseless and helpless human beings caught between internal and external enemies was beyond belief."[214] It was the French, and not the Mainz citizens, who forced the sufferers to return to the city.

On July 22, a ceasefire finally took effect. The final capitulation followed the next day, at which point the Mainzers could freely ride into the city. The announcement of ceasefire and freedom was met "with loud approval" by citizens. From then on, the Mainz Jacobins were considered collaborators. When caught, they were humiliated and abused. In *A Voyage round the World*, Forster had said that "a proper sense of the general rights of mankind awaking in them, will bring on a revolution." By contrast, the Mainzers returning from exile were now the ones to perpetrate violence against the "Clubbists." At first, Goethe interprets it as nothing but the "righteous anger of these people who had been so unbelievably insulted and humiliated" toward the despotic apostles of freedom. Then, he became witness to an "archClubbist" being beaten: "All his bones are broken, his face is unrecognizable."[215] The Mainz republic was history.

As an observer, the experience for Goethe was about the driving forces of history, which is to say, it was about violence and destruction. The state of martial law nearly overpowered his faculties—"I am unable to think," he writes, as a way of describing the inconceivable.[216] He openly admits to having been eager to experience "cannon fever" firsthand, through his own immediate experience, rather than from hearsay. He is afflicted by it in the battles around Mainz.[217] But while accompanying the field campaign against France in 1792 (which ended with the Cannonade of Valmy, giving the French a decisive victory), he rode to the front lines, risking his own life, to expose himself to what it was like to come under fire—and putting embedded journalism to the test. While under martial law, his form of witness turned inward to self-observation:

Even under these circumstances, however, I could soon tell that something unusual was happening inside me; I paid careful attention to this, and yet I can only describe this feeling in the form of a simile. It seemed as thought I were in a very hot place, and thoroughly permeated by that same heat, so that I felt completely at one with the element in which I found myself. My eyes lost nothing of their strength or clarity of vision, yet it was as though the world had a certain reddish brown tone, which made my own condition, as well as the objects around me, still more ominous. I did not notice any quickening of the blood[;] everything seemed rather to be caught up in this intense burning. From this it became apparent to me in which sense this condition could be called a "fever." Nonetheless, it is still remarkable how whatever causes such terrible anxiety is conveyed to us solely through our sense of hearing; for it is the thunder of the cannon, and the howling, whistling, and blasting of cannon balls through the air that is actually the cause of these sensations.[218]

Forster's and Goethe's views converge on this one point more than any other: the world must be experienced and felt. Both willfully approach the extreme, whether the fever of revolution or the fever of war for the defeat of the republic. Both seek to understand the nature of these events. On this one point, they are intimates, transcending all differences. Goethe would have felt this more than Forster. Beyond the chaos of the revolution, he kept faith with him.

V

The End

The Great Perplexity, 1793–1794

:: :: :: :: ::

Forster arrives in Paris on March 29, 1793. The next day he gives a speech before the French National Assembly, in which he makes the case for the accession of the Mainz Republic to France. The request is sustained. Forster's return to Mainz is made impossible by the occupation of Mainz by Prussian-Austrian coalition troops and the capitulation of the city on July 23, 1793. While his wife and children seek safety in Straßburg, he remains in exile in Paris. In the meantime, the French Revolution enters the terror phase. Thousands of political opponents and alleged counterrevolutionary activists are guillotined. In November Forster secretly meets his children, his wife, and her lover—and later husband—Ferdinand Huber, in Switzerland one last time. In December, after finally setting to work on new manuscripts about the political environment in Paris, Forster's health worsens dramatically. On January 10, 1794, Forster dies in a garret in the Rue de Moulins in Paris.

:: :: :: :: ::

The Monstrous Head of Revolution: Paris

"Have you heard anything from Forster?" Goethe asks toward the end of 1793 in a letter to Soemmerring. He seemed to have vanished off the face of the earth. After his arrival in Paris on March 20 of that year, nearly all his letters are addressed to his wife. Contact with everyone else breaks off rather abruptly. Three days after Forster left Mainz, the French commander-in-chief, Adam-Philippe de Custine, was driven out of Bingen and Kreuznach by antirevolutionary coalition troops. His forced retreat to Landau makes it impossible for Forster and his companions to return to Mainz. This changes everything. Forster has just six shirts and the coat on his back, he writes in a letter to his wife. It was going to deal a heavy blow for his exile to be prolonged: "I would miss these poor few possessions less than the loss of my writings, my drawings, and my laboriously scraped-together books."[1]

As a political rebel, Forster falls under Kaiser Franz II's decree that prohibits all German subjects from collaborating with the French Revolution's regime. The Imperial Ban also threatened Forster with treason, the verdict for which was a heavy penalty to be decided by the Kaiser with approval by the Reichstag; it was to be applied against enemies of the Reich in conjunction with the declared martial law. Ostracized traitors forfeited all privileges, freedoms, and rights; their possessions were confiscated.

Horst Dippel has meticulously traced the Imperial Ban as having been "never actually, neither de jure nor de facto, imposed on Georg Forster, nor another Mainz Clubbist." Although publicly threatened by the Kaiser, the required confirmation never cleared the Reichstag. Regardless of its legal validity, the Mainz Jacobins viewed themselves as being subject to the Imperial Ban. In all, five hundred club members were arrested. Even if legally it was not put into effect, the Imperial Ban can be considered effective in its stigmatizing power. Hence, Forster could not set foot on German soil without facing arrest. He was told how public opinion was turning against him: "The Berlin intellectuals argue about me, throughout Germany I am foully spoken of, I pass for the chief instigator of all that is rotten in Mainz."[2] A precarious situation, even for the friends he left behind.

There were certainly attempts at loyalty beyond all political fault lines. Albrecht Schöne calls attention to how the members of the Göttingen Society of Sciences responded to this situation.[3] Forster was a member, as was the Jacobin Philipp Friedrich von Dietrich, both of whom were henceforth considered unwelcome—at least by the imperial government, which, in April 1793, demanded their debarment from the Göttingen Society. Forster's father-in-law Heyne was secretary of the academic society. He drew up a position paper rejecting the request for debarment as "*beneath all dignity* of the Society" and circulated it among the society's members. One could "expect such a thing from courtiers and sycophants, but not a corpus of scholars." Because their society was "an academic society, not a political corpus, nor a club." Heyne was selective in distinguishing between scholarly and political engagement: "Whatever is happening with regard to the political ties and relationships of our members has no bearing on this society; because such affairs have no bearing on academic and scientific matters." He concluded with a refusal to carry out the debarment of both members Forster and Dietrich: "If the imperial government wants to have the men in question stricken off, so it may recommend it; and if it wants it to be made publicly known, so it will be announced, as happening at the command of the imperial government."[4]

Forster's political activities certainly did not find unanimous approval among the society's members. Some recorded their displeasure in writing. Nevertheless, all signed Heyne's statement. Even Forster's friend from his Göttingen days signed: "I likewise Lichtenberg." Forster and Dietrich were not debarred.[5]

But attempts such as this to protect Forster from the stigma of treason were the exception. With the capitulation of Mainz on July 23, Forster's fate was sealed. He persevered in his Paris exile, virtually destitute and cut off from his friends in Germany. He lacked connections in Paris: "Not a soul benefits from my stay here. I am completely isolated, I know nobody." It was as if he had dropped out of existence. His crash was bottomless; no net caught him. "I know," he wrote his wife, "that I am merely fate's ball now, but I am affected all the same by where it tosses me. I have no home, no fatherland, no kinsfolk anymore; everyone who was otherwise attached to me has gone on to form other connections, and when I think of the past, and think of myself as having had attachments, it is my own choice to do so, my own way of thinking, and not a necessary result of my circumstances."[6] Forster has maneuvered himself into the margins.

His letters from Paris could be viewed as self-pitying. Yet they document how Forster picks himself back up again and again, pushing himself to work on his manuscripts and continuing to draw up plans. His discipline, which carried him his whole life long, was now becoming brittle, though, and the unruly pain of having lost everything was beginning to seep through the cracks of his existence. "Today as I was walking alone in the *palais royale*," he confessed, "my eyes spontaneously filled with tears at the thought of returning to my room. In this endlessly large city, I have no one who worries about me in the least, no one who has taken an interest in me, and if I were to disappear tomorrow, there is no one who would not be completely indifferent!" Wherever he goes, loneliness receives him—"I cannot reconcile myself to the dead solitude, and yet, I still despise it less than the sorry company of people." His life was slipping through his fingers. In a letter to his father-in-law of December 10, 1791, Forster surmised that he still had "a little over ten years" left[7]—when it was actually only a few months.

Those few months were among the darkest of his life. First, there was the personal misfortune. Separated from his wife and children, he describes his grief as a "limitless, incurable misery." His marriage had failed, his wife was proceeding with plans for divorce, and Forster became resigned:

> I cannot live with you, yet I cannot do without you; it is impossible that I
> will ever be happy through love, because never can another object stir me
> and fill my heart so—and so absolutely and devotedly did I love! I still love,
> even with cleaved consciousness! Never, never to have been happy, never
> to have kindled requited feelings, consequently, never to awaken. Do not
> wish to fathom this hell, but wish that I learn to accept what I deserve, so
> that I may die with greater calm and reconciliation to my fate. I was certainly made for domestic happiness, I was of use as a man, and could have
> become ever more so, as a man, as a father and friend, as a husband. All is
> ruined.[8]

Forster has arrived in the "port of resignation." He is tired and weary of life: "I am at peace, but I am burned out."[9] The loss of his two children pained him especially—to be able to be with them "for one hour," he said, "I would chop off one of my fingers if it would buy an hour." He does not permit himself to think of them: "This memory kills me."[10]

In addition to the collapse of his private life, he was shocked by the political reality that he found in Paris. If Mainz was the center of developments in Germany, now he was in the center of the revolutionary world. Paris was the "monstrous head" of world politics.[11] What a disappointment! Just days after his arrival, he wrote his wife: "From a distance, everything looks different than it does up close." The French nation remained foreign to him; it was "frivolous and volatile, without stability, without warmth, without love, without truth, led by the head and not the heart, by fantasy and not feeling."[12] He had expected revolutionary fervor and instead found a "quixotic fanaticism" resulting from an ideological brainchild. Initiated into the "secrets of local intrigue," familiar with the "revolting labyrinth in which everything here twists and turns," a terrible suspicion comes over him: "[It was] an absurdity to have sacrificed my last ounce of strength and worked with honest zeal for a thing that no one else genuinely means, but which serves as a mere cloak for raging passions."[13]

He was successful in his mission before the National Convention. On March 30 he gave a brief speech, in which he argued: "Les Allemands libres qui Vous demandent la reunion, sont ambitieux de partager la gloire qui attend le nom française"[14] (The free German who wants to belong to France will strive to share the glory expected of the French reputation). The National Convention approved the accession of the Mainz Republic—why not?—but quickly became beset by larger problems than the fate of what was ultimately a small territory—and one that was as far from Paris as anything can be said to be far from Paris.

Once again Forster felt the revolution entering a critical phase. "Everything is now on the cusp," he wrote his wife.[15] "Everything here tilts toward decision." George Danton, supporter of the radical La Montagne party, successfully declared himself in favor of the Revolution's establishing a tribunal, which was launched on March 10. Together with Jean-Paul Marat and Maximilien de Robespierre, Danton introduced the Revolution's phase of terror—invoked by the catchword *la Terreur*. The tribunal's role was to sentence political opponents and perform executions. This terror was not the result of fanatical individuals acting alone or outside of the law, but the result of state policy that was passed for the alleged protection of the Revolution. For Forster, the French nation was spurred to its most "severe paroxysms of the revolution."[16] Who was interested in the affairs of the Mainz Republic, where the fate of world history had come to a bloody resolution?

It only took a few days for Forster's faith in the Revolution to fall away. "You wish for me to write the history of this abominable time?" Forster asks, responding to an idea of his wife's. "I cannot do it. O, ever since I learnt that there is no virtue in revolution, it sickens me. Far from all idealistic reverie, I could go to destinations with imperfect people, fall on the way, stand back up, and keep going. But with devils, and heartless devils, as they all are here, it is a sin against humanity for me, against the sacred mother earth, and against the light of the sun." Bitterly he must have realized that he had placed all his prospects in what now proved to be a spurious cause. "If ten months ago, or eight months ago, I knew what I know now, I would *without any doubt* have gone to H or A," to the liberal Hamburg or Altona, "and not to the Jacobin Club." He now distances himself from the Revolution as rashly as he first threw his support behind it: "With every day and every hour I am more convinced that my political career has ended." Nevertheless, he continues his political writing during the months he is in Paris. In addition to "Darstellung der Revolution in Mainz" (Account of the revolution in Mainz), Forster works on an essay entitled "Über die Beziehung der Staatskunst auf das Glück der Menschheit" (On the relationship of statecraft to human happiness) as well as a series of narrative sketches of Paris. Yet, all these works are subject to the most bitter provision that Forster is capable of as a writer: "I write what I no longer believe."[17]

What remains are plans, which never materialize. Forster stands before the wreckage of his life. Cut off from all income in Germany, he lives "hand to mouth."[18] At least the National Convention grants him a payment of eighteen livres daily.[19] By comparison, a printer of the *Encyclopédie* of Diderot and d'Alembert earned two livres per day, a foreman three, and a learned editor four. A cheap edition of the *Encyclopédie* cost ten livres, whereas a horse cost one hundred.[20] Forster is humbled: "After so many years of strenuous work, everything that I have undertaken to get ahead has failed, and from now on I begin the world anew, as it were, without knowing how, as I am cut off from all of Europe, overwhelmed by debt, lacking any means or support here, and nearly without prospect." The offer by a German newspaper to collaborate while in exile proves to be a "miserable stopgap." He wants to find work as an independent writer again and thinks "a voyage to another part of the world would be the best for me in order to reestablish my inner peace as a naturalist by giving me a practical occupation."[21] He even begins learning Persian. He has a destination in mind, too: "To India!"[22] At least these daydreams revitalized him: "If I had my things from Mainz here, I would be glad! I would take with me to India a travel library and my paint box and my nice drawing paper." Ways and means alone were lacking. Indeed, "twenty friends of liberty" would have to be found in England, each of whom was to donate a little money "for one man, who has sacrificed (and erroneously, as it turns out) his entire civic existence for freedom."[23] Yet, who would offer patronage to someone who had wagered everything he had on such a risky political game?

Nature consoled him, as it so often did, at least temporarily. Forster saw "the pleasure of the first green on the trees" for the last time, and he found a garden where he would walk alone at daybreak "while everyone is still sleeping," serenaded "by the hundred nightingales[;] how beautiful is nature, when one does not think, does not remember, but rather merely lives in the present moment." It is his receptivity to nature that "even on the saddest of days" provides him with "an invaluable number of happy moments in the purest enjoyment of nature": when we are "open to its impressions, nothing beautiful escapes us, nothing great, nothing good, nothing stirring in the universe without our heartstrings sounding of it."[24] As Forster wrote this, the executioners were reporting for duty at the guillotine on the Place de la Révolution.

The Cold Fever of Terror

"The heads fell like roofing tiles," Antoine Quentin Fouquier-Tinville remarked on the conditions in Paris.[25] From March 1793 to August 1794, 16,594 people were sentenced to death in France. The archives list 2,639 executions in Paris alone, as the capital of the Revolution. When one factors in the executions of those suspected of opposing the Revolution but who had been denied due process, as in Nantes, Lyon, or Toulon, the estimated death toll climbs to 35,000–40,000.[26] The number of arrests was even higher at a half million persons. People of all social classes were affected. No one, according to the historian François Furet, escaped the "collective tremors."[27]

The cold logic of the reign of terror blindsided Forster. He was horrified to observe state-legitimated terror. There were no misguided fanatics at work here, no individuals who were above the law—terror was the law. It was elevated to a maxim of state. Robespierre found a formula for describing revolution as "the war of liberty against its enemies." This was to be taken literally, and the slightest deviation from revolutionary doctrine was enough to be viewed as enmity. The alleged virtuousness of terror only fed its resoluteness. For revolutionary power to count, according to Robespierre, "if the mainspring of popular government in peacetime is virtue, its resource during a revolution is at one and the same time virtue and terror: virtue, without which terror is merely terrible; terror, without which virtue is simply powerless. Terror is nothing but prompt, severe, inflexible justice; it is therefore an emanation of virtue."[28]

Just one month after his arrival in Paris, Forster expressed resignation, in a letter to his wife, about his chances of gaining a foothold while in exile:

"Intellectual merits and the talents of businessmen are worth nothing now. Whoever swims to the top sits at the helm only until he is ousted by the next strongest person a moment later. When a person cannot pursue anything but can be denounced and guillotined, that person is nothing." He saw the violence, and in particular the death penalty, as an unavoidable means of revolution. He was in no way, as is sometimes claimed, "opposed to violence." On January 21, 1793, Ludwig XVI, the last king of the *Ancien Régime*, was publicly beheaded. It was not the first time a death sentence had been carried out against a king—Karl I had been condemned for high treason and put to death in England in 1649. But that was a long time ago—who in Paris would have been thinking of that at the sight of the beheaded Ludwig XVI? Instead, Europe held its breath as the judgment was carried out against royal leaders and common citizens alike and the thread of continuity based on monarchist rule was severed after centuries of believing it to be God-given. Forster expressly welcomed this execution. For him, it was not a conviction "according to the law books, but rather the laws of nature," because Ludwig had offended the rights of humanity. He openly expressed his approval of this deed, which shook up the history of the monarchy. Soemmerring could only shake his head in disapproval to Heyne: "Such childishness, to let his mustache grow as a sign of satisfaction with the execution of the king, I never would have suspected him of this"[29]—the French commander Custine was known as Général Moustache for his mustache and was thus the model for this gesture of solidarity.

Now it was different, though. Forster recited to himself and others the formulas that seemed to justify the excess of violence in Paris: he believed in the "*importance* of the Revolution in the larger cycles of human destiny"; he said, "[I am] still satisfied with the Revolution, *even if it is something completely different than most people think*"; and "in order to be able to bring about that which is *good*, you must let it all and even more happen, you must even let that which is *bad* happen."[30] For Forster the revolution is still a natural event that cannot be judged by human measure. And since nature "does not seem moral at all," the fate of individuals remains "always secondary to the general fate of the entire species."[31] This makes for an inscrutable future: "We must patiently wait to see what happens, because fate has set its own instruments in motion: incalculably passionate people."[32] To avoid being liable to "bleak fatalism," one must not doubt "that the effects of blind, irrational power are weighed against the plans of the whole, and are woven in such a way that their discordant notes are lost in the general harmony."[33] Revolutionary rhetoric put into neutral.

Still, radicalism flares up in Forster's willingness to view the unthinkable as natural: France must "swim in blood and in tears," and one must "not be surprised by anything more now. Everything is possible." Personal happiness is worth nothing. "Thousands and thousands of families may perish, but our great work will not back down anymore." Forster can still work himself into a rage as an apocalyptic prophet of the Revolution: "It *should* be this way! Our century should end with catastrophe. Long will the balance sway back and forth perhaps, individual people will come to be regarded as nothing in the violent clashes, but reason and equality will prevail. It is already being decided here, even now. No one's death and execution will cause any greater a sensation because of their name or title, or because they are of a certain rank, and that is the proper point." The guillotine makes everyone equal. It is "a fate that watches over us."[34]

Yet, his correspondence increasingly struck a different tone, which marked a turning away from his latent readiness for violence. Moderation is scarcely possible in revolutionary times; he says, however, "I also do not like the resolve, which hunts through thick and thin with these inveterate devils." Forster is most concerned by the "cold fever" that seizes people. Perhaps it can only be helped by being cut out like a cancer. "But thank god I am not a miracle doctor—I could not do it, as much as I see the necessity."[35] Forster considers himself to be a passionate Jacobin as well as a revolutionary of liberty, but only in his imagination does he cross the boundary to using violence.

Dancing on the Brink of Absurdity: Adam Lux

On July 13, 1793, Jean-Paul Marat was stabbed to death by Charlotte Corday. Her conviction did not take long: on July 17 she was guillotined on the Place de la Révolution—today the Place de la Concorde—in Paris. Marat was one of the most radical revolutionaries, an agitator who sought to whip up public opinion with his newspaper *L'ami du peuple*. Along with Danton and Robespierre, he put a violent face on the Revolution. To purge the Revolution of the sordidness of its opponents, Marat, "the great reaper of men" was intent on carrying out mass killings in grand style.[36] At the beginning of September 1792, nearly twelve hundred prisoners in Paris prisons were murdered, people who had not necessarily been arrested for political reasons but who served as symbols for all enemies of the Revolution. Should it be unclear who was ultimately responsible for this bloodlust, it was the triumvirate of the top three revolutionaries. Charlotte Corday was twenty-four years old when she was awoken to take a stance by putting an end to the reign of *Terreur* through assassination. The virtue of terror had become lost to the Revolution; Corday's execution exposed that fact when it was turned into a scandal: the executioner, before the eyes of everyone there, slapped the face of her severed head. He was indeed out of line, having violated the rules of propriety that applied even to those sentenced to death.

Forster was an eyewitness at this execution. He was impressed by the calm with which Corday went to her death—and by her exceptional beauty. Forster, terribly ill and deathly tired, wrote his wife, "She was radiant with health, tantalizingly beautiful, mostly because of the unspoiled charm that hovered about her. Her dark-brown hair was cut short, making for an antique head on the loveliest bust. She persisted in her serenity until the last moment on the scaffold, where I saw her executed. It seemed to me that her death went well for her. 'You suffered briefly,' I thought." She had the makings of an icon of resistance against the terror of the revolution, as Forster saw it, invoking the "purity of her soul." "She loved the republic and liberty with enthusiasm and deeply felt her inner breakdown. Her memory lives among hundreds of thousands who still have a sense of simple greatness."[37] He could think this, even write it in private correspondence, but he could not say it publicly.

Forster was accompanied to Paris by Adam Lux, who had written a dissertation on enthusiasm years earlier at the university in Mainz. Now, as one of the delegates of the Mainz Convention in Paris, he was to present the request for the young republic's unification with France.[38] Lux was even more deeply impressed by Charlotte Corday—he was enraptured. On the way to the scaffold, their eyes met: it was "the gaze of an angel, which penetrated deep in my heart, and moved it to intense stirrings," he gushed. Forster wrote his wife that Lux had "completely lost his head over this girl."[39] This might not be given much import if the death of the martyr for liberty had not resolved Lux in his own plans to emulate her by committing political suicide. He wanted to die in the name of liberty under threat, even though his wife and children would be left unprovided for in Mainz. Forster became witness to his fanaticizing, which he was appalled to observe but was unable to stop.

Lux knew what he was doing when he published a vindication of Marat's executed murderer. On July 13, in his pamphlet *Avis aux citoyens français*, he warned of danger to the French fatherland when vice was triumphing everywhere. In letters, he announced that he hoped his own death would "incite a crisis of some furor among citizens." An example must be set: "I believe that my death will be of greater use than my life, and will bring some sensation to the lethargy which prevails." He described his hopeless determination to his wife: "Liberty cannot win, so I am lost in every way and unable to continue."[40] For liberty to be victorious, sacrifice was necessary. Nevertheless, at this time his suicide was merely an announcement; although his July 13 pamphlet brought him trouble, it did not immediately bring him to the scaffold.

Two days after the execution of Marat's murderer, Lux distributed the leaflet *Charlotte Corday*. Her assassination was politically honorable, because she had indeed saved the lives of many thousands of people by murdering Marat. Lux attested to Corday's "heavenly soul,"[41] which yearned for liberty and love of fatherland. She should be memorialized. Then he accused the governors of the Revolution of having "deceived the French people": "I sought the Reich of gentle liberty here, but instead I found the suppression of merit and virtue, the triumph of ignorance and crime. I am tired of living for so long under all these abominations, which you carry out, and looking even further at all the misfortune that you prepare for the fatherland! Just two hopes are left for me: either to suffer as a sacrifice for liberty and die on this honorable scaffold, or to contribute to dispelling your lies" and "putting an end to your tyranny."[42]

Forster was stunned. He did not expect a public attack on revolutionary practices in Paris. Forster proved to be cautious. "My fellow deputy, who is still with me," he reports to his wife, "has given free rein to the vehemence of his feelings and put his opinions about the events of the day in print. He invites upon himself the indignation and perhaps the vengefulness of those who are capable of anything. His intention is noble, his courage heroic, his feeling right and good, but with the overall prevailing *mistrust*, there is a risk of compromising those who have no part in his proceedings, who even feel and think differently, and who are convinced of a different destiny. I am meanwhile prepared for all of it." Forster senses danger for his associate— and for himself. Calmly he explained to his wife, when all was said and done: "I was always more frightened for him than for me."[43]

As was bound to happen, Lux was arrested. All was not lost, though. "Poor quixotic Lux will probably have to sit for a while until he seems harmless," Forster reckoned. But Lux, as his posthumous defender Alfred Börckel wrote a century later, had planned his suicide "in cold blood."[44] Every amnesty, every loophole for escape—his friends suggested non compos mentis—Lux brusquely rejected.

The revolutionary tribunal sentenced him to death. On November 4, 1793, at five o'clock in the evening, he was beheaded on the same plaza where Charlotte Corday had been. His friend Georg Kerner reported that he did not relent on his way to the guillotine, "calling out the names of tyrants to the few people whom he encountered on his way. He ascended the scaffold like a rostra." Other witnesses reported that he set foot on "the scaffold with an almost improbable cold-bloodedness. He spoke, smiled, and faced death without fear. On the scaffold, he embraced his executioner before he met the blade."[45]

Forster was not there. The news reached him in Pontarlier a few days later on November 10. The "unfortunate Lux," he wrote, became "a martyr for liberty on the guillotine, in accordance with his wishes. This news has ruined my whole day." The gap between Forster and his former crony is unmistakable. Few are "so committed to their ideas that they give their lives for them as Lux did."[46] What are Forster's "ideas"? Nothing but pipe dreams, for which life is to be wantonly sacrificed. He does not deny Lux any recognition for having "bravely achieved his martyrdom for Charlotte Corday. He must have stood before the tribunal quite at ease and said that he knew he was guilty according to the laws of death and would gladly suffer it." But shaking his head, he adds: "He *leapt* onto the scaffold."[47]

Finally, Forster understands that the time of revolution has gone mad in foolhardy ways. In revolutionary times, "the least probable becomes the most probable."[48] The Revolution becomes uncanny to him. Charlotte Corday may have been fanatical; it is unclear whether her actions were taken "on the basis of error or truth," but she went about the deed in a gracefully disconcerting way: "She guided the knife with certainty, without having practiced."[49] But Adam Lux was a deluded idealist, in Forster's view. The Revolution had swallowed him. Forster recognizes an abyss opening up before him. It cannot be contained. "It means dancing on the brink of absurdity, does it not?"[50]

Back to Nature: Human Dignity

During those months of Forster's greatest personal misfortune, separated from his wife and children, he forced himself to work on what he thought would be the conclusion to his essay "Darstellung der Revolution in Mainz" (Account of the revolution in Mainz): a new essay titled "Über die Beziehung der Staatskunst auf das Glück der Menschheit" (On the relationship of statecraft to human happiness). Forster had complained since arriving in Paris that he could "no longer find the peace of mind essential to work,"[51] and indeed, the text is written with a noticeable trace of final elan. As a seismographic document of Forster's shock over the Paris events, it is quite remarkable. Forster builds upon nature's curative strength one last time, by seeking an alternative, an exit from the labyrinth of terror.

This late contribution can also be read as an incipient retreat from all political hope, a weary recoiling from the daily turmoil of concrete politics, which Forster observes with increasing distance. It had long ceased to be about concrete claims for him. Regardless, Forster is a marginal figure of little influence in Paris. Weary of the political reality, he seems to take some license in committing himself to the luster of a noble idea that might be called "natural humanism," thereby invoking the "true friend of humanity."[52]

Summoned before the judgment seat of reason, as Forster begins his argument, it would be an enthralling spectacle to see government and politics account for their consequences, if they were to defend "against the charge of virtue, and against the testimony of experience." His depiction of politics writ large outside of France is devastating. "The happiness of humanity, according to the assertion of rulers, is the unchanging aim of their governing concerns," he writes of the self-perception by rulers, so that he can subsequently wrest the mask of human philanthropy from Europe's despotic states.[53] The actual "secret to all state policy" runs counter to any professed care for their subjects: "*expansionism*" and the "secret to all politics, *cunning*, and *misanthropy*. Yet, secret I say! In our times, the courts scarcely cloak their intents in such veils anymore; only the means for realization, the machinery and its gears, remain covered until the opportune time."[54] The mechanisms of power are worth uncovering to analyze their impact. The "unshakable strength of the internal state machinery" produces people with a "marionette nature," because despotism requires "automatons"[55] and thus a mass that "moves machine-like according to prescribed laws" within the sphere of activity and is accustomed to "viewing their leaders and teachers as beings of a higher form, as miracle workers and gods. In order to be consistent, despotism must want this *moral nullity* of humanity."[56]

These allegations were not new and were not always brought forth by Forster, but they are characteristic of how rulers were commonly chastised. You can almost feel the boredom of repetition creeping into the handy formula. For Forster, though, this is merely the springboard for broad criticism, which Schiller anticipated in his 1795 societal critique *The Aesthetic Education of Man*. What came to be called alienation in the twentieth century, Forster calls "Verstümmelung," or mutilation. Despotism is, at its core, misanthropic. It makes "people poorer than nature created them."[57]

The development of one's abilities is served by a humanity that is oriented toward nature and bears in mind "what people can and should be." The constitution of a state has this purpose to serve, because providing a "lasting form of constitution and of moral formation, which does not detract from individuals' free use of all their powers, but merely determines what society must demand of its members, while granting them the inestimable advantages of personal and property security, which each person honors because he is endowed with a perfectibility, through which his own purpose for himself must be honored—this does not seem to lie entirely outside of the realm of possibilities."[58]

Forster would have been familiar by then with at least the premise of Kant's moral philosophy. He adopts the formulation that a person must never be used merely as a means, but rather must always be respected as a purpose unto himself, too. On the tailwind of this Enlightenment imperative, Forster can declare a time when "that false image of *happiness*, which for so long has stood as the goal of the human life course, goes crashing from its pedestal, and the true guide of life, human dignity, is set in its place."[59]

In contrast to "mutilation" by despotism, the diversity of natural gifts is to be respected and fostered. Nature is "less step-motherly than its detractors would portray it," and "the *tired* worker is not always too obtuse to think; the joy of hard-won gains opens the gates of receptivity for him, too."[60] Forster devises an almost natural human upbringing. Because instead of squandering time with "foolish masquerades, prattled formulas, vulgar drivel about incomprehensible things, boring lessons in barren knowledge," it would suffice "to direct the common man's attention to himself and his circumstances, to incite his thirst for truth, and to awaken in him the desire to be and to become, by his own effort, that for which nature has called him in his particular state and endowments into existence. Communicating a more useful, more applicable knowledge of nature leads to one's own reflection, and through this, the revival of that delicate sense, which makes us a party to *reasonable* joy; this nice concern by philanthropists demands neither exceptional gifts, nor outsized strengths."[61]

It is already there in each one of us. It must only be developed. Forster formulates a political swansong that would have pleased Goethe: "What makes man virtuous and happy, no government and no education can give him; *it is in him*, but can all too easily be stifled by the malice of tyrants and the doting love of the teacher!" Nature becomes his source of recovery one last time. "Instead of promising us happiness," Forster shouts to the rulers of the world, "your sole concern should be to clear away the barriers that obstruct the free development of our powers; open up the path for us, and we will walk it, without the help of your drover's prod, to the destination of moral education; then, you will see! We will receive *joy* and *sorrow*, our true teachers, from mother nature's hand!"[62]

All of this is directed against the misanthropes, the despots, the tyrants and rulers of the world. Yet, there is a peculiar lack of addressee in the text. Cut off from his German audience in Paris, Forster hopes Voss will print the text in Berlin because "the tone should offend so few." Yet he speaks as though from nowhere, it seems. He was at no time in his life idealistic enough to believe in the efficacy of a reasonable admonishment. He remained far more loyal to his view that people "of that sort of idealistic perfection, which we so often dream of in books, are practically nonexistent."[63] Instead, he insists on a "respect for the rights of others that has yet to take place in the real world," because the world is "not yet ripe for it."[64] Why, then, the talk of human rights and natural development?

The text can be read on two levels. Or, to put it another way: if one reads it in the opposite direction, it becomes conclusive. Forster did not have in mind the rulers of Europe, who would not have listened to him anyway, so much as the governors of the Revolution in Paris. His outrage at the reign of terror precluded any idea of personal happiness in any event, but his defense of a human dignity that is to be respected is like an intellectually constructed wall against violence. And since reason alone is not capable of accomplishing anything, nature is invoked, the salubrious effects of which are subsequently developed.

After the execution of his comrade-in-arms Adam Lux, Forster became aware of just how thin the ice was. When addressing despots, his considerations remain broad and politically correct. In April of that year, he described his impressions of Paris in a private letter to his wife, in which he spoke of the "*tyranny* of reason" as perhaps "the staunchest of all" to still face the world.[65] This is to say, the most dangerous form of despotism does not come from the old decrepit European state, but from the fanatically Enlightened Revolution that is fueled by reason and placed at the center of the political world.

Upon arriving in Paris, Forster described the Revolution as "one of the great means of fate," which must not be "contemplated in relation to human happiness and misfortune."[66] Individuals cannot be taken into consideration. The madness of his experiences in Paris had, by all accounts, caused him to see the necessity of finding a bulwark against all political expectations— through the proclaimed dignity of humanity. "Never before has tyranny had such brazenness, such exuberance, never have all principles been so trampled, never has calumny prevailed with such rampant brutality."[67] This is from a letter to his wife—in reference not to the despots of the world but to the revolutionaries in Paris. His noble idea of respecting the dignity of humankind owes itself not to flights of reason, but to the overwhelming circumstances of his day.

His proposal for a return to nature may sound desperate. Because is it not nature which revolutionary violence fatefully advances through the passions of its protagonists? The same nature that also created reason, whose tyranny now poses a threat? Here at the end of his life, Forster holds the loose threads of his thinking in his hands. They cannot be connected anymore.

The Revolution Is the Revolution

In his last piece of writing, which Forster worked on in December of 1793 and set aside just a few weeks before his death, he mounts one last defense of the French Revolution for his reluctant German readers. From the courage that comes with desperation, he counters every defamatory and critical objection. Hostility toward the Revolution had only become louder: in 1790 Edmund Burke published *Reflections on the Revolution in France and on the Proceedings in Certain Societies in London relative to that Event: In a letter intended to have been sent to a Gentleman in Paris*; in 1793 the influential German translation by Friedrich Gentz was published. Burke was a critical opponent of the French Revolution. In *Reflections on the Revolution in France*, he rails against the "Apostles of Liberty," leveling the accusation against France that "everything seems out of nature in this strange chaos of levity

and ferocity. And of all sorts of crimes jumbled together with all sorts of fol-
lies." He defends the nobility and clergy against the "hue and cry" and desires a "manly, moral, regulated liberty."[68] Burke's allegations are fueled by
his earliest impressions of the terror encompassing Paris.

Forster, for his part, was now writing one letter after another—which
were posthumously published as "Parisische Umrisse" (Parisian sketches,
hereafter referred to as "Parisian Sketches"). Joachim Heinrich Campe had
already published *Letters from Paris at the Time of the Revolution* in 1790 in
order to portray the events from his own immediate experience for the German public. But Forster did not act like a reporter in any true sense. His observations hover above events without calling things directly by name, so to
speak. The guillotine is not mentioned once.

The common interpretation is that Forster wrote "Parisian Sketches"
from the perspective of a "propagandist and hermeneutist of the Revolution."[69] He persevered not only "in his voyage to the capital of the revolution but also in his positive image of the French Revolution, characterized
by enthusiasm and belief in reason through concrete-empirical experience
and observation."[70] This is simply out of the question. While Forster was
writing "Parisian Sketches," he admitted, under the guise of confidentiality,
what a great tragedy it was that his "enthusiasm *de sa belle mort*" had died.[71]

It is equally untrue that Forster pursued the strategy of "refusing experience," as he was alleged to have done, in light of the state terror purge
in Paris during those months. Resisting experience in favor of preserving
ideological hopes disqualified him as a critical observer of the Revolution's
events. If anything, his immediate experience with state-legitimated mass
killings was what began distancing him from the practice of revolution. To
recall a motto that hangs over Forster's late work: "I write what I no longer
believe."[72]

Although Forster seeks to defend the Revolution in "Parisian
Sketches"—he does not want to abruptly let go of his fascination with current world history when he is still trying to make sense of it—his outline of
the features of the Paris Revolution is cryptic. This text, too, can be read on
two levels. Forster did well not to let his underlying meaning openly emerge
amid a heated political situation. Despite the drone of revolutionary slo-
gans and expressions about human progress, in "Parisian Sketches" Forster
draws a decisive outcome from his experiences: the Revolution cannot be
comprehended.

Forster begins by pointing out that the capital of France has for a long time been "the lofty school of human knowledge"—an allusion to the venerable Paris university. "Now more than ever," he continues, "it requires only a short visit and a fleeting glance here to become awake to what would take decades elsewhere to even imagine, and not only to decipher the *spirit of the present* but the *signs of the future*, too." Paris is the center of the political world, the departure point for the revolutionary age. It is the "monstrous head" that steers everything.[73] In short, "*Paris* sets the tone." And then Forster reaches for a comparison that could be considered a brilliant flourish if he were trying to flatter the Parisians: "Paris is to the new republic what Rome once was to the empire"[74]—a subject that will be returned to.

Forster praises the Parisian Revolution in quite politically correct terms: "It is the greatest, most important, most astonishing *Revolution of moral education and development in the whole human race.*"[75] Against the decadence of a noble caste, it enforces a liberating "simplicity in mores" and conquers pernicious luxury, even banishing "silver spoons from the table."[76] The Revolution is the lofty school for the fraternity—or better put, the partnership—of humanity. Forster is optimistic that the Revolution, "as a work of providence, is exactly in the right place of this noble plan for the education of the human race."[77]

Why, then, the violence, the terror, the great tremors? According to Forster, the events possess an "unstoppable momentum," which is in line with the "nature of revolution"[78] and which, "as a motive force, is not purely intellectual, not purely rational," but is "the raw force of the multitude."[79] The Revolution rolls like a wave, comprising a mass of individuals who cannot be stopped. If the individual perishes, the nation moves on. It is apparent "that without some hard struggle, it will not turn out."[80]

Primed by his overall thinking, Forster elevates this natural fatalism to a breathtaking defense of terror. The Revolution's savagery and coldbloodedness, according to Forster, resemble the yoke of despotism, since both involve unrestrained violence. However, the "injustice loses its shock, its violence, its arbitrariness, when public opinion decides to pay homage to the law of necessity as absolute arbiter in the last resort, which gives rise to that action or ordinance or measure." The implications of this sentence, which Forster has the audacity to write, are ghastly. "Universal reason," in the sense of "*public opinion*," has to follow the law of nature, the law of necessity, and since nature is neither moral nor immoral, appalling injustices lose their unlawfulness if they are natural. Thus, political terror is turned into a natural phenomenon that is beyond good and evil. It is natural "that without streams of blood, the advantages of the revolution, which the world needs so desperately, would not have come to pass." In all the horror, the "unfathomable and inscrutable wisdom of providence" steers the revolution.[81] The idea of a "Theodicy" of the Revolution,[82] which Forster shies away from working out in any detail in "Parisian Sketches," is a consistent expression of his understanding of revolution as being natural. As spectacular as his ideas are for us to read, to him they are conventional.

It could be tempting for the reader at this point—weary of revolutionary ideology—to set aside "Parisian Sketches," but that would be premature. It is important to pay attention to the cracks in the text, the imperceptible fragility in the argumentation, the contradictions that pervade "Parisian Sketches" and reveal Forster's fumbling to make sense of his egregious experiences. The text vibrates beneath the surface with fear of the absolute ruthlessness of the revolution. "We have sacrificed ourselves for our entire species, or, it is equally valid to say, we have let ourselves be sacrificed," he says with bitter resignation about the revolutionaries. One would like to "doubt whether the revolution is made more for the people than the people are made for the revolution."[83] When Forster wrote that Paris was the "monstrous head" of the revolution, the ambiguity of his phrasing now comes to the fore. The Revolution is the monster that threatens to devour everything. Reason cannot cope with it, nor can an idealistic history of philosophy: "Is there even a considered, calm, firm course of reason to be found in this scheme?" Forster asks. "Or is it everywhere passion against passion and dice against dice?"[84] Below the surface of world historical catastrophe, Hegel would still see reason at work, whereas Forster is not willing to ascribe reason to the development of the Revolution. He is obstinately not an idealist.

If he starts out "Parisian Sketches" by praising the Revolution as marking the progress of the moral education and development of the entire human race, just pages later he depicts a political apocalypse, in which people are inferior creatures to the revolutionary powers of nature. If the "revolution was decided in the council of the gods,"[85] how does it concern human beings? The "total impression" of the Revolution becomes "so colossal" that it disrupts the measure of man.[86] "When the bomb bursts, will it not shatter everything?"[87]

With the "colossus" motif, Forster takes up the antiquating moment of his earlier comparison with Rome. He has already given the reader a hint about the way in which "Parisian Sketches" could be read against the grain. Surely it was quite conventional to compare Paris in terms of its world historical significance for political modernity with Rome's place in antiquity. Appearances are deceptive, though. Forster's reference to Rome is the Trojan horse of "Parisian Sketches"—equipped, of course, with well-hidden contents that no reader could discern. Ostensibly, Forster only means to confer the laurel of world historical significance on Paris. An incidental remark contains the sole key to interpreting his private correspondence. "Do you remember the description of the Roman empire in the first volume of Gibbon's?" Forster wrote his wife at the end of June 1793, before he began "Parisian Sketches" but after his upsurge of unsettling experiences in Paris. He was referring to a voluminous historical work: between 1776 and 1788, Edward Gibbons's *History of the Decline and Fall of the Roman Empire* appeared in London in a six-volume set. Forster had praised it in his "History of English Literature from the year 1788." He points his wife expressly to the chapter in which Rome falls prey to the Praetorian Guard—Caesar's bodyguards. After the death of the tyrant Caligula, as Gibbon writes, the consul attempted to establish a free republic under the watchword "freedom." While they were deliberating, the Praetorians took over: "The dream of liberty was at an end; and the senate awoke to all the horrors of inevitable servitude." In just a few sentences, we have the blueprint for an interpretation of France's liberation from despotism, its attempt at republicanism, and the destructive violence of its ideologues. The ancient "tumult of fermenting parts of the world,"[88] as Forster puts it, quite resembles the modern.

Forster may have had Gibbon's Roman history in mind when he wrote his wife from Paris: "It looks here now as it did then in Rome. Never before has tyranny had such brazenness, such exuberance, never have all principles been so trampled, never has calumny prevailed with such rampant brutality." A signal to the epochal downfall can be found in "Parisian Sketches" beneath the surface of Forster's apologia for the revolutionary conditions. The Revolution resembles a global conflagration. Perhaps "a god batters the human species to pieces in order to cast it anew in the crucible?"[89]

The Revolution cannot be comprehended. It escapes human understanding, and it depends—as Forster puts it in "Parisian Sketches"—"on a higher order of things." Forster turns into an eschatologist of the political: "All of this is so much greater, following a much wider-ranging plan." Upon consideration of the forces opposing the Revolution, Forster touches on "the secret workshops of fate, which scoff at all calculation, at a higher instance of world government, which destroys human activity through men."

The Revolution is our destiny. It is a fate that befalls us. "Its fulfillment is not of this world. I learn this more each day," Forster wrote in early 1793.[90]

Not even nature's knowledge is enough to decipher the revolution. Forster wanted to find a law of revolution in "the political phenomena of one moment," because revolutions "perhaps also have their cycles."[91] Now he must acknowledge that a law of politics writ large cannot be found. In one of his darkest statements, Forster draws consequences from this realization by letting all understanding hit the tautological mark and offering the fictional recipient of his letter a definition: "The revolution is—provided that you are lusting after our generalized definition—is the *Revolution*." This sentence permits two readings: first, Forster conceives of the phenomenon of political revolution as the true revolution of his epoch, since political upheaval had come to be viewed merely as "new inexorable momentum."[92] Political modernity created revolutions—Forster recognized this and designated it as his epochal signature. Thus, it is the second entry that he emphasizes in his tautological formulation.

But Forster uses this formula to fulfill the desire for a "generalized definition," too. He does so while offering a definition that explains nothing but offers affirmation through interpolation: "The revolution is . . . the *Revolution*." It resembles the Old Testament revelation of God, "I am who I am," which may have still been in Forster's ear from his father's sermons. Thus, the revolution for him becomes an event that can only be experienced but not thought about. It would be "madness" to "want to put a stop to the revolution or to erect boundary markers," he points out, with not just a political but a rational attempt at containment in mind: "A natural phenomenon"—the subject is still the revolution—"which is too rare for us to know its peculiar laws, cannot be restricted and determined by the rules of reason, but must maintain its free rein."[93] Although people will try to settle into the maelstrom of events, it in no way means taking the path of "the revolution's incalculable torrent," mapped out by reason. "All morality," Forster now states in opposition to all Enlightenment ideals, "seems to me a farce and a vulgar invention, which we have for one another's benefit. To the detriment of human powers, nothing is capable of changing the fate of the entire race, nor the fate of a single individual. Everything will be unavoidably snatched away, to suffer and to cause suffering, until the elasticity wears out and snaps."[94]

The revolution is the revolution—there's nothing else to be said.

Forsaken Like a Child

The end is bitter. When someone like Georg Forster "burns and wears out the body at both ends," when his own body remembers the hardship of a voyage around the world as a chronicle of maladies, when he is "wholly enfeebled and skeletal" and ultimately finds himself "the most lonesome man in the world," as Klaus Harpprecht describes him, when his familial happiness is fractured and his hope for the Revolution has gone rancid—then he goes on living while having disappeared from the consciousness of others. He is now nothing more than "a silhouette,"[95] Forster says of himself, a specter, a shadow.

At the beginning of November, Forster travels from Paris to Pontarlier on the French border. In the small Swiss village Travers, he meets Therese, his two children, and Huber for the last time. They have not seen one another in an entire year. Therese must not set foot on French soil. She is German and therefore an enemy of the French. She cannot go to Germany, either, because there she is considered a Jacobin and the wife of a leading revolutionary. Forster is extremely cautious. Their meeting is brief and is kept secret. Their farewell is bleak. "I mustered all I could to hold it in, but now it bursts forth. O my children! How my heart bleeds at this farewell! You will have seen what very happy hours I have spent with you. It is sad for me to face the coming weeks and perhaps months." He is almost defiant in his willingness to plant himself against all adversity in life, but he swiftly meets his threshold after returning to Paris. "I have helped myself the best I can, but everyone has their limit."[96]

His first collapse occurred in early December. After going out "without an overcoat, at night, in a nasty Paris fog," he was bedridden, convulsive, sleepless. He did not recover; brief periods of respite were only followed by further relapses. The days passed like leaden weights. "A whole hour for me to get dressed and shaved, and now I lie here like a fly in my armchair."[97] He could barely walk a hundred steps. Fortunately, he was cared for by Johann Georg Kerner, a journalist and physician from Swabia, and he had the help of other doctors and a Polish assistant. Forster was given opium and underwent "sad, lonesome, long nights . . . exhausted from the day, unable to write nor read, and yet [he needed to] stay up in order not to spend an even sadder, longer night sleeplessly in bed!" In soliloquy he addresses the specters of death, training his sights on the "incorrigible injustices of nature," yet he compels himself to carry on: "You must, until you cannot any more, then, the end will come on its own." All glory fades. The man who sailed around the world becomes a shadow of himself, his former greatness and importance waning: "Every wretched dog can die."[98] Yet he pulls himself together enough to write a handful of letters to his wife and children, later said to be "the saddest in the German language."[99]

He still spoke of politics, because the volcano of the Revolution did not remain silent. "The earth still shudders beneath our feet, the ground still burns." But as if in a delirium, Forster's thinking becomes short of breath and disjointed. He only occasionally spouts watchwords, after letting go of the manuscript for "Parisian Sketches." As though to spur him in his radicalism, he reaches—"God be with us!"—for Machiavelli's *The Prince*. Its justification for violence and barbarism would have interested him. The Revolution, he jots down on paper, is a "hurricane—who can stop it? A man who is brought to action by it does things which can only be understood in posterity, not at the moment of direness. But the slant of justice is too high here for mortals. What happens *must* happen." At the very end, in his last letter, he writes as if delirious: "We have won everywhere, just like lions." But it has been a long time since he has understood what is happening. Nature and revolution cannot be thought of in the same terms anymore—"so my hypothesis takes to the lifeboats," as he once remarked.[100]

When he first arrived in Paris in April of that year, he confessed to his wife how forlorn he felt. "For the first time in my life," he wrote, "none of my contrivances can help me with anything, and I stand as forsaken as a child, who has no means to feed itself." Now he will find himself "in the days to come . . . crying like a child for myself alone, so deeply fatigued am I." The pain is so severe that he goes "mad from one moment to the next."[101]

On December 20 he leaves his garret in the Rue des Moulines, near today's Opéra (Palais Garnier), one last time. On his way back, he cannot find a carriage and has to walk across half the city to get home. He does not recover from this. For ten days, he does not close his eyes. He cannot hold a quill anymore, writing only a few last lines on January 4, 1794, to his children. A few words are better than nothing. Forster, who wrote constantly all his life, falls silent. "I now have no more strength to write. Farewell! Guard yourselves against illness; kiss my sweetheart."[102] Six days later, early in the evening of January 10, he dies.

A Source of Strange Introspection

"You will pardon me a question coming from a place of true respect and confidence," Sophie von La Roche cautiously writes in a letter to Forster's friend Soemmerring on January 29, 1794. "Is it true that our Forster, loved so long by us in Paris, has died?" The news of his death gradually made the rounds—and was met with reticence. On January 18 an obituary appeared in the Paris *Gazette nationale*: "A bad fever, sequelae from his voyage at sea, domestic woes, and his work may have blown him off course. But nothing shook his ardent love for the revolution; his final wishes were for the republic and his children." To Germans, he was a rebel, an outlaw, a political enemy. Who now would admit to remembering him?[103] There were private avowals of friendship, for instance, by his father-in-law Heyne: "I loved the man inexpressibly; he meant more to me than a child." But public statements were dangerous. "O how I would have liked to dedicate a few pages to him," Lichtenberg writes, himself now a family man. "Would that I were still the free-thinking and free-writing being that I used to be, when I was childless and unencumbered by the future. Now free *thinking* must be left at that."[104]

It was different for Goethe. Often chided as the prince of serfs, he alone possessed the authority to establish a memorial. Klaus Harpprecht references the civic idyll *Herrmann und Dorothea*, which Goethe began writing in 1796 as an approving discussion of "enthusiastic liberty and laudable equality."[105] The epic depicts a young man who

at the first fire of lofty thoughts, strives for noble liberty
Takes himself to Paris, and soon meets a horrible death.[106]

Goethe, who regarded himself as a man of action, esteemed the "love of liberty" and the "desire to affect in new, changed ways."[107] Forster is commonly identified with this character. However, was Forster actually being portrayed here, or is it Adam Lux who serves far more as the template (the character remains unnamed)?[108] Biographical accuracy cannot be expected of Goethe. So the young man is driven to Paris only to find there "Kerker and death," which does not apply to Forster, since only Lux was imprisoned in Kerker. However, at another point in a letter to Soemmerring, Goethe writes that Forster "had to atone for his mistakes with his life,"[109] which, on the level of documentary accuracy, is not in keeping with Forster's end.

Moreover, *Herrmann und Dorothea* speaks of the "bridegroom's death" with regard to the young man; Lux was already married when he set out for Paris. Goethe's creative liberties cannot be subsequently tamed by philological meticulousness. Nevertheless, the image of the young man in Goethe's epic poem likely depicts, as Nicholas Boyle concludes, "a character like Adam Lux."[110]

The true memorial that Goethe endowed to Forster is found in a seemingly isolated but significance place. Goethe knew that he and Forster agreed that sometimes on the world stage one has to move "over a bed of hot coals."[111] Forster may have lost his way in Goethe's eyes, too, although Goethe still kept faith with Forster and thought of Forster's loyalty when defending his understanding of nature against Newton. It was all or nothing for Goethe. In the historical part of *Theory of Colours*, which he considered his most important work, Goethe thanks Forster by name for his support.[112] In this way, too, loyalty to an outlaw can be professed.

Apart from this prominent example, the political disgrace into which Forster had fallen meant that most people shied away from any form of public memorial. Forster swiftly became a forgotten author. He could have been granted the rank of a classic writer of German prose, but the antirevolutionary and reactionary ethos of the long nineteenth century, which sought to suppress demands for political freedom, caused his impact to quickly fade. Heinrich Heine, himself driven into exile in Paris, explicitly remembered his poor predecessors, who "had to atone most bitterly for their revolutionary sympathy." Indeed, those who fled to Paris "disappeared or died here in poverty and misery"—and he adds: "I also saw not long since the garret in which citizen Georg Forster died."[113]

There were further, albeit sporadic, apologists, such as Albert Leitzman, who, in his 1891 inaugural lecture "Ein Bild aus dem Geistesleben des achtzehnten Jahrhunderts" (A picture from eighteenth-century intellectual life), sought to ease the fixation on Forster's revolutionary years: "With bitter pain and moral loathing, he realized he was wrong, losing his faith not only in the revolution, but in his life's purpose, which he now regarded as unsuccessful, while never giving up his keen belief in the ideal tasks of the state, in the future of humanity and in the possibility of their realization." It was of little use. The accusation prevailed that, as Friedrich Christian Laukhard puts it, "particularly Georg Forster's heated *Afterpolitik*"—a state policy without religion and morality—"was supremely guilty of depraving so many."[114]

As a travel writer, Forster gradually found his way back to his audience—a work like *A Voyage round the World* cannot be ignored forever. Alexander von Humboldt explicitly emphasizes Forster's achievement in his comprehensive physical description of the world, the five-volume *Cosmos*. His "distinguished teacher and friend" pointed the way to natural description like no one else in the German language: "Gifted with refined aesthetic feeling, and retaining the fresh and living pictures with which Tahiti and the other fortunate islands of the Pacific had filled his imagination . . . George [*sic*] Forster was the first gracefully and pleasingly to depict the different gradations of vegetation, the relations of climate, and the various articles of food, in their bearing on the habits and manners of different tribes according to their differences of race and of previous habitation. All that can give truth, individuality, and graphic distinctness to the representation of an exotic nature, is united in his writings."[115]

Although Humboldt never hid his affinity to Forster's republican views, Forster's rehabilitation as a travel writer and the renewed favor of his audience came at the cost of ignoring Forster's politics. It is a "shame" to this day that Forster, "because of a brave decision made out of political conviction, which put all of his German contemporaries to shame for its historical impact," does not occupy his due rank among the philosophical theorists of revolution. Hannah Arendt's significant book *On Revolution* contains only three lines about him. And in *The Structural Transformation of the Public Sphere*, Jürgen Habermas grants a single mention on half a printed page to Forster, who fought for a political public.[116] There may be reason to disregard Forster's significance to the understanding of revolution in that Habermas—representative of many—assumes a philosophical self-conception of modernity's civil revolutions. In Hegel's line of interpretation, philosophy is the original stimulus for revolution. John Locke, Rousseau, Thomas Paine, and others become the initiators of revolutionary upheaval. In this Habermasian interpretation, which is part of the determination of relationships in *Theory and Practice*, Forster is not once named.[117]

The particular appeal of Forster's thought and deeds lies in his unwieldiness with respect to the Enlightenment's common optimism of progress. While he stressed the naturalness of revolution against the primacy of reason, he crossed over to the dark side of the Enlightenment. He gave priority to immediate experience above reason, action before theory. "The composure with which we judge worldly affairs from our sofas or our desks, and soon absolve the parties involved, soon condemn them, ceases at the site of action; we stand there, over a bed of hot coals, as it were, and hearken to the omnipotence of those conditions which have been preparing this inexorable fate for centuries."[118] Forster falls out of step with the self-conception of Enlightened political modernity.

Still, his failure is illuminating—not only for us, but for Forster himself—as is his late perplexity, by not being able to make any natural sense of politics writ large anymore. "It is strange," Forster writes, summing up his life since the start of the Mainz Revolution, after all political and personal hopes had slipped through his fingers in Paris,

> that our own peculiar circumstances are so tied up in the most important matters of the entire human race! When I even consider how little everything I have done since November seems expedient now, I sometimes almost wish I had quietly left Mainz and settled in Hamburg or Altona, where I would have nothing to do with people's affairs. When I consider by contrast, that ultimately a certain development of myself was only possible in this way, which although infinitely painful, has also become a source of strange introspection for me . . . then, I am content with all that has happened.[119]

One does not have to agree with Forster, nor share his theory of the naturalness of revolution—nor endorse the primacy of immediate experience above all theory—to be able to locate a "source of strange introspection" in a life so rich with experience. His career represents the most remarkable display of the "development of an endowment" that sought to garner "more life and rigorous reality" for the "awareness of the whole" of nature[120]—to the point of political action and at the price of self-sacrifice.

Epilogue

The Mahogany Trunk

Forster's life knew few constants. He took hasty steps through just four decades, traversing great change. Among the few possessions that accompanied him throughout his life was a small mahogany trunk that he used as a writing desk. He purchased it just days before boarding the *Resolution* in June 1772. Forster kept this small trunk with him during his voyage to Antarctica and the South Seas. It crossed the equator with him, survived frost, scorching heat, and dampness on board during many a storm, and returned intact to Europe. It served as an indispensable tool in Forster's workshop at nearly every station of his life, from England to Kassel and from Wilna to Mainz. It was still in his estate after his death, among the manuscripts he had tucked away and left behind in Mainz. Beyond its tangible function, this small mahogany trunk contains a metaphorical richness in its symbolism of a life spent in the perseverance of writing. In all his alacrity for experience, in all of the various locations in which he worked, in all his active enthusiasm for revolution, Forster was a lifelong writer, for whom the page provided self-assurance and dialogue with his readership.

The expressiveness in his language lifted him above every rupture—in all his books, in his essays and journal entries, in speeches and reviews, in his translations, too, and finally in the large number of his letters. His life may have fallen to pieces, but the array of his experiences is preserved by the coherence of his writing. And his language shines on.

Notes

Prologue

1. G. C. Lichtenberg to J. F. Blumenbach, April 23, 1791, in Lichtenberg, *Schriften und Briefe* (Munich: W. Promies, 1967), 4:791.

2. G. Forster, "Ueber das Verhältniß der Mainzer gegen die Franken," AA X/I, 23.

3. G. Forster, *Cook, the Discoverer*, 258.

4. For a biography of Cook, see F. McLynn, *Captain Cook: Master of the Seas* (New Haven, CT: Yale University Press, 2011).

5. Ludwig Uhlig rightfully called for a look at the "whole" Forster, Forster having stood "quite at the center of German intellectual life." Uhlig, "Georg Forster und seine Zeitgenossen," 155. Nevertheless, the voyage around the world and the Mainz Republic have marked the two poles of his biography, which can only be linked through an appropriate attention to all phases of his life.

6. Herder, "Reflections on History of Greece," 186–187.

7. Herder, "Ideas for Philosophy of History," 291.

8. Rousseau, "Social Contract," 210; Montesquieu, "Du l'esprit des lois," in Montesquieu, *Oeuvres complètes* (Paris: Bibliothèque de la Pléiade, 2001), 2:558: "Plusieurs choses gouvernent les hommes: Le climat, la religion, les lois, les maximes du gouvernement, les exemples des choses passées, les moeurs, les manières; d'ou il se forme un esprit général qui un résulte."

9. G. Forster, *Cook, the Discoverer*, 160.

10. G. Forster, "Ein Blick in das Ganze der Natur. Einleitung zu Anfangs-gründen der Thiergeschichte," AA VIII, 78.

11. G. Forster to C. G. Heyne, May 5, 1785, AA XIV, 319.

12. G. Forster, "Versuch einer Naturgeschichte des Menschen. Anlagen in Menschen," AA VIII, 158.

13. G. Forster, "Parisische Umrisse," AA X/I, 593.

14. G. Forster, "Versuch einer Naturgeschichte des Menschen. Anlagen in Menschen," AA VIII, 159.

15. Tilo Schabert traced the connection between the conception of nature and the French Revolution but did not mention Forster in this book, although he had been living in Paris since late March 1793 and had commented on Forster's political position, for example as expressed in his work "Parisische Umrisse." Schabert, *Nature und Revolution.*

16. Baxmann, *Die Feste der Französischen Revolution*, 9.

17. Lepenies, "Historisierung der Natur und Entmoralisierung," 265.

18. S. Ascher, *Ideen zur natürlichen Geschichte der politischen Revolution, ohne Ort 1802* (Kronberg/Ts.: 1975), 4, 81, 78.

19. E. Burke, *Reflections on the Revolution*, 469.

20. Ibid., 456–457.

21. H. C. Harten and E. Harten, *Die Versöhnung mit der Natur. Gärten, Freiheitsbäume, republikanische Wälder, heilige Berge und Tugendsparks in der Französchen Revolution* (Reinbeck bei Hamburg, 1989), 112; for the figure of trees planted, see 110.

22. Joubert, *Carnet*, 361.

23. G. Forster, "Erinnerungen aus dem Jahr 1790 in historischen Gemälden und Bildnissen von D. Chodowiecki, D. Berger. Cl. Kohl, J.F. Bolt und J.S. Ringck," AA VIII, 267.

24. G. Forster to T. Forster and L. F. Huber, December 28, 1793, AA XVII, 498.

25. Goethe to Schiller, March 9, 1802, in *Correspondence between Schiller and Goethe, from 1794–1805*, trans. L. Dora Schmitz (London: G. Bell, 1877–1890), 407.

26. Forster's wife, Therese Forster (later Huber), relayed this remark by Johann Reinhold Forster in a letter to C. G. Heyne dated January 16–22, 1794. Huber, *Briefe*, 285.

27. Goethe to S. T. Soemmerring, February 17, 1794, in Goethe, *Werke* (Weimar), section 4, 10:142.

28. See these biographies, which do, however, contain such accounts: Saine, *Georg Forster*; G. Steiner, *Georg Forster*; Harpprecht, *Georg Forster*; L. Uhlig, *Georg Forster*. U. Enzensberger's book *Georg Forster* is also essential.

29. Stilett, *Von der Lust*, 24.

30. G. Forster, "Darstellung der Revolution in Mainz," AA X/I, 511.

31. G. Forster, "Fragment einer Vorrede," in *Neue Beyträge zur Völker- und Länderkunde*, part 1, 1790, AA V, 377.

32. G. Forster, "Leitfaden zu einer Geschichte der Menschheit," AA VIII, 192.

33. The extensive bibliographies on Forster account for an ongoing, if not growing, reception for his work since the Akademie edition of his complete works and since the inception of Georg Forster studies by the Georg Forster Society. However, Forster was still not able to take his rightful place in a commemorative culture that was cross-disciplinary, public, and nonacademic. Horst Dippel has rightly pointed out that "the

development of a commemorative culture around Forster that transcends one milieu has never been achieved." This "parceling out," as he calls it, has still led to a prevailing fragmentation in the reception of Forster's works. Although Forster research has produced a wealth of individual results, attempts at a complete account are in the minority. It will take a comprehensive view of the individual results and, above all, allowing the "whole Forster" to speak by prioritizing source material over secondary literature. See Dippel, "Georg Forster in der deutschen Erinnerungskultur," 28.

34. G. Forster to T. Forster, August 1, 1793, AA XVII, 409.

35. G. Forster to F. H. Jacobi, January 2, 1789, AA XV, 233.

36. See Ewert, *Vernunft, Gefühl und Phantasie*. Ewert lucidly works out in his study the appeal of essay-writing for Forster's thinking, which consists in enabling him to reflect without requiring a system: "The literary form of the essay suspends the phenomenon of a philosophical approach without dedicating itself to the abstractness of a school of philosophy" (64). The essay furthermore allows for the "treatment and tolerance of contradictions" (9).

37. Ralph-Rainer Wuthenow highlighted the emancipatory aspect of the essay: Wuthenow, *Vernunft und Republik*, 55:

His aim is not to influence . . . persuade and win, but to purely instruct and educate an audience, out of respect for each reader. This reader is not criticized for expanding his knowledge; rather, he is offered a form of representation that appeals to his maturity and freedom, and at the same time, is the medium for their development. . . . Thus, in the essayistic form, it is not just content that is conveyed as the answer to a question about the nature of enlightenment; the essayistic approach is itself a response in its open, dialogic, affable character: the essay appears here as an essentially liberating genre that imparts maturity.

38. F. Schiller to L. F. Huber, January 13, 1790, in Schiller, *Werke*, 25:391.

39. G. Forster, *Ansichten vom Niederrhein, von Brabant, Flandern, Holland, England und Frankreich, im April, Mai und Junius 1790*, AA IX, 116.

40. G. Forster to T. Forster, April 25, 1790, AA XVI, 107.

41. G. Forster, "Vorlesungen über allgemeine Naturkenntnis," AA VI/2, 1754.

I

1. Cf. J. Locke, *An Essay concerning Human Understanding* (Oxford: Publisher, 1975), 104. Here Locke says he conceives of the mind as "white Paper, void of all Characters, without any Ideas."

2. G. Forster to J. G. Herder, January 21, 1787, AA XIV, 623.

3. G. Forster to J. K. Spener, February 12, 1785, AA XIV, 279.

4. Unfortunately, it cannot be determined which books in his father's library Georg Forster had access to. Regina Mahlke verifies some of them in "Georg Forster und die Bibliothek."

5. For a biography of Reinhold Forster, cf. Hoare, *Tactless Philosopher*.

6. G. Forster, "Can the world ever reasonably and through reason be happy?," AA VIII, 360.

7. G. Forster to T. Heyne, August 1, 1784, AA XIV, 143.

8. Ibid., June 23, 1785, AA XIV, 345.

9. G. Forster to F. H. Jacobi, December 20, 1783, AA XIII, 514; G. Forster to Herder, July 21, 1786, AA XIV, 513; Heine quoted in Eliot, "German Wit," 136.

10. G. Forster, "Ein Blick in das Ganze der Natur. Einleitung zu Anfangsgründen der Thiergeschichte," AA VIII, 90 f.

11. G. Forster, *Ansichten vom Niederrhein*, AA IX, 128.

12. G. Forster, *Cook, the Discoverer*, 237.

13. G. Forster, "Something More," 147–148.

14. G. Forster, "Der Brodbaum," AA VI/1, 70.

15. G. Forster, *Voyage round the World* (2000), xxxv.

16. G. Forster to T. Heyne, August 1, 1784, AA XIV, 145.

17. Uhlig, "Theoretical or Conjectural History," 407, 404 (quote).

18. My characterization of the young Forster as a blank slate may nevertheless be found to be an exaggeration. Critical to that assessment, however, are not the identifiably early influences of philosophical-scientific schools and readings, but Forster's ability to dispense with that form of conditioning in favor of an impartial view. How astonishing his openness was toward those impressions during the voyage—when he encountered cannibalism, sexual permissiveness, or a lack of awareness of property—may be read in contrast to his resentment toward the Poles. An affirmative tendency allows him a perspective that is nearly free of judgment, at least as a guiding principle; where this was lacking, he fell back on a corset of conventional judgments typical of his time. On the significant difference between these perspectives in Forster, cf. Fagot, "Polen als Georg Forsters Gegenstück."

19. G. Forster, *Ansichten vom Niederrhein*, AA IX, 128.

20. Cf. Dalos, *Geschichte der Russlanddeutschen*, 12–33.

21. G. Forster to J. G. v. Zimmerman, December 30, 1787, AA XV, 82 f.

22. J. R. Forster, "Ueber Georg Forster."

23. Cf. the corresponding remark by Reinhold Forster in his review of the book *Enchiridion historiae naturali inserviens* on March 27, 1788, AA XI, 498.

24. G. Forster, "New Holland."

25. For Forster's stay in England, cf. Gordon, "Reinhold and Forster in England."

26. G. C. Lichtenberg to E. G. Baldinger, January 10, 1775, in *Lichtenberg's Visits to England, as Described in His Letters and Diaries*, trans. B. L. Mare and W. H. Quarrell (Oxford: Clarendon Press, 1938), 63–65.

27. Benjamin, "Paris."

28. Cf. Gordon, "Reinhold and Forster in England," 42–46.

29. J. R. Forster, "Ueber Georg Forster," Stück 16, April 15, Spalte 124.

30. J. R. Forster to T. Pennant, October 3, 1768, cited by Uhlig, *Georg Forster*, 35.

31. I. Kant and H. S. Reiss, "Conjectural Beginning of Human History," in Kant, *Political Writings*, 233.

32. G. Forster, "Über historische Glaubwürdigkeit," AA VII, 37.

33. Cf. Gordon, "Reinhold and Forster in England," 58 f.; Dippel, "Georg Forster und England."

34. For the history and significance of the Royal Society, cf. Gribbin, *The Fellowship*, 125.

35. G. Forster to Jacobi, August 29, 1783, AA XIII, 474.

II

1. G. Forster, *Voyage round the World* (2000), 631.

2. Ibid., 378, 374.

3. G. Forster, *Cook, the Discoverer*, 159. For Forster's reflected Eurocentrism and tendential cultural relativism, cf. Braun, *"Nichts Menschliches soll mir fremd sein."* Jörg Esleben emphatically points to the significance of Forster's lifelong translation work as an act of intercultural communication in his "Übersetzung als interkulturellen Kommunikation": "Forster conceives of translation as the essential element of all dimensions of intercultural communication. . . . The external perspective makes translation for Forster an important means for attaining a differentiated image of other cultures. Above all the inclusion of those perspectives made possible by translation is indispensable" (178).

4. G. Forster, *Cook, the Discoverer*, 157; G. Forster, *Ansichten vom Niederrhein*, AA IX, 86.

5. G. Forster, *Cook, the Discoverer*, 159.

6. G. Forster to T. Forster, April 14, 1790, AA XVI, 80.

7. Ibid., September 25, 1793, AA VXII, 448.

8. G. Forster, *Voyage round the World* (2000), 10; G. Forster, *Cook, the Discoverer*, 242.

9. J. W. v. Goethe to C. v. Stein, December 7, 1781, in Goethe, *Werke* (Weimar), section 4, 5:232; G. Forster to T. Heyne, July 25, 1784, AA XIV, 141.

10. G. Forster to J. K. P. Spener, September 16, 1782, AA XIII, 400.

11. G. Forster, "Vorlesungen über allgemeine Naturkenntnis," AA VI/2, 1755.

12. G. Forster, *Voyage round the World* (2000), 9, 10.

13. G. Forster to Spener, September 17, 1776, AA XIII, 53.

14. G. Forster, *Voyage round the World* (2000), 6. For Forster's claim of wanting to produce a "philosophical history of travel," cf. Neumann, "Philosophische Nachrichten."

15. G. Forster, *Cook, the Discoverer*, 242. "Even if the scientific facts, the results of fieldwork, are often in the foreground," writes Gerhard Steiner, summarizing Forster's primary interest, "man stands in the middle-ground of descriptions and trains of thought." Steiner, *Georg Forster*, 18.

16. G. Forster, *Voyage round the World* (2000), 9–10.

17. G. Forster to F. H. Jacobi, August 29, 1783, AA XIII, 474.

18. G. Forster, *Voyage round the World* (2000), 9.

19. Ibid., 7; G. Forster to Spener, January 14, 1779, AA XIII, 173; Goethe, *Poet and the Age*, 546.

20. G. Forster to Spener, April 2, 1776, AA XIII, 32.

21. G. Forster to C. G. Heyne, January 28, 1792, AA XVII, 36.

22. G. Forster to T. Forster, April 16, 1793, AA XVII, 345.

23. G. Forster, "Über lokale und allgemeine Bildung," AA VII, 54.

24. G. Forster, "Über historische Glaubwürdigkeit," AA VII, 32; G. Forster to C. G. Heyne, August 30, 1790, AA XVI, 177; G. Forster, "Über die Insel Madagaskar," AA V, 625 f.; G. Forster, *Cook, the Discoverer*, 173.

25. G. Forster to Spener, September 16, 1782, AA XIII, 402, 399.

26. G. Forster to F. A. Vollpracht, December 31, 1776, AA XIII, 77.

27. Ludwig Uhlig rightly points out the desideratum for a comparison of the English and German versions of Forster's account of the circumnavigation. Out of consideration for the German reading public—whose needs, according to Forster's own assessment, were "quite markedly different" from the English readership's (foreword to the translation of *Des Capitain Jacob Cook dritte Entdeckungs-Reise*, AAV, 187)—Uhlig's opinion is that the German version of *Voyage round the World* is often so "free and ambiguous" with respect to the English original that it prompts misunderstandings. Uhlig, "Der polyglotte Forster," 137. For the breakdown of the relationship between the German and English, see also Peitsch, "Georg Forsters 'deutsche' Kommentierung."

28. G. Forster to Spener, December 27, 1776, AA XIII, 75, 76.

29. Ibid., July 19, 1781, AA XIII, 332. The remark is in reference to his work on the translation of his father's *Observations Made during a Voyage round the World*.

30. G. Forster to Spener, August 25, 1783, AA XIII, 469; G. Forster to Jacobi, January 22, 1789, AA XV, 252.

31. G. Forster to Jacobi, September 18, 1790, AA XVI, 191; January 24, 1789, AA XV, 255; G. Forster to J. R. Forster, September 8, 1783, AA XIII, 476 f.; G. Forster to Jacobi, November 1, 1789, AA XV, 363 f.

32. G. Forster to J. H. Campe, October 5, 1779, AA XIII, 242; G. Forster to Jacobi, September 18, 1790, AA XVI, 190; G. Forster to Spener, September 16, 1782, AA XIII, 399. This is not a moment of disgruntlement. In a letter to J. G. von Zimmerman, March 13, 1788, Forster writes about his general dependence, saying that he "belongs to that class of people who unfortunately must sell themselves in order to make a living." AA XV, 128.

33. G. Forster to J. H. Merck, November 13, 1783, AA XIII, 504; G. Forster to F. H. Jacobi, August 29, 1783, AA XIII, 474; G. Forster to Merck, November 13, 1783, AA XIII, 504.

34. H. Blumenberg, *Shipwreck with Spectator: Paradigm of a Metaphor for Existence* (Cambridge: MIT Press, 1997), 7–8.

35. See Braudel, Duby, and Aymard, *Die Welt des Mittelmeeres*.

36. G. Forster, *Voyage round the World* (2000), 19, 653, 19.

37. G. Forster, "Reise von London nach Paris 1777," *Tagebücher*, AA XII, 7.

38. G. Forster, *Voyage round the World* (2000), 18.

39. Ibid., 293, 292.

40. Ibid., 266–267.

41. Ibid., 75, 618, 354.

42. Ibid., 619.

43. Ibid., 497.

44. Ibid., 500.

45. Ibid., 111.

46. Ibid., 634.

47. G. Forster to T. Forster, April 14, 1790, AA XVI, 80; G. Forster, *Ansichten vom Niederrhein*, AA IX, 236.

48. G. Forster to T. Forster, April 14, 1790, AA XVI, 80.

49. According to Braudel, *Das Mittelmeer und die mediterrane Welt*, 27.

50. Ibid., 145.

51. G. Forster, *Voyage round the World* (2000), 61, 693.

52. Ibid., 655.

53. Ibid., 43, 653, 39.

54. Ibid., 298, 78, 672.

55. Ibid., 594, 292.

56. Ibid., 370; G. Forster, *Reise um die Welt*, ed. G. Steiner (Berlin: Akademie-Verlag, 1989), 414; G. Forster, *Voyage round the World* (2000), 286.

57. G. Forster, *Voyage round the World* (2000), 684.

58. G. Forster to Spener, September 19, 1775, AA XIII, 21.

59. G. Forster, *Voyage round the World* (2000), 598.

60. G. Forster, "Reise von Kassel nach Wilna 1784," *Tagebücher*, AA XII, 122.

61. G. Forster, *Cook, the Discoverer*, 202–203.

62. By Hauser-Schäublin's account, "Das Leben an Bord."

63. G. Forster, *Cook, the Discoverer*, 218; G. Forster, *Voyage round the World* (2000), 293, 295; G. Forster, *Reise um die Welt* (1989), 379.

64. G. Forster, *Cook, the Discoverer*, 234; G. Forster, *Voyage round the World* (2000), 62.

65. G. Forster, *Cook, the Discoverer*, 219.

66. Cf. Bucher, *Die Spur des Abendsterns*.

67. Cook, *Journals of Captain Cook*, vol. 2, *Voyage of Resolution and Adventure*, 323: "I . . . had Ambition not only to go farther than any one had done before, but as far as it was possible for man to go."

68. G. Forster, *Voyage round the World* (2000), 597.

69. Ibid., 70, 76.

70. Ibid., 297.

71. Ibid., 291, 649, 350, 351–352.

72. J. R. Forster, *Observations During a Voyage*, 69.

73. G. Forster, *Voyage round the World* (2000), 64.

74. Ibid., 68, 69.

75. Ibid., 76.

76. Ibid., 68.

77. Ibid., 71.

78. G. Forster, *Cook, the Discoverer*, 184; G. Forster, *Voyage round the World* (2000), 76.

79. G. Forster, *Cook, the Discoverer*, 184–185.

80. G. Forster, *Voyage round the World* (2000), 288.

81. Cook, *Journals of Captain Cook*, 2:305: "Our situation requires more misses than we can expect."

82. G. Forster, *Voyage round the World* (2000), 291.

83. Ibid., 67.

84. Ibid., 68.

85. J. R. Forster, *Resolution Journal*, 198.

86. G. Forster, *Voyage round the World* (2000), 292.

87. Ibid., 293.

88. Ibid., 291.

89. Ibid., 78.

90. Ibid.

91. Cook, *Journals of Captain Cook*, 2:323: "We could not proceed one Inch further to the South"; G. Forster, *Voyage round the World* (2000), 651.

92. J. R. Forster, *Beobachtungen während der Cookschen Weltumsegelung*, 8; Cook, *Journals of Captain Cook*, 2:198.

93. See Uhlig, *Georg Forster*, 60.

94. G. Forster, *Voyage round the World* (2000), 141–142.

95. Ibid., 91.

96. Ibid., 298; G. Forster, *Reise um die Welt* (1989), 125.

97. Cf. Joppien, "Georg Forster und William Hodges."

98. G. Forster, *Voyage round the World* (2000), 143.

99. Cf. Garber, "Reise nach Arkadien."

100. Bougainville, *Voyage round the World*, 106.

101. G. Forster, *Voyage round the World* (2000), 80, 160; Horace, *Satires, Epistles, and Ars Poetica*; G. Forster, "Antwort an die Göttingischen Recensenten," AA IV, 54.

102. G. Forster to J. G. Herder, July 21, 1786, AA XIV, 512; Börner, *Auf der Suche*, 142. See also Veit, "Topik einer besseren Welt"; Sangmeister, "Das Feenland der Phantasie."

103. G. Forster, *Voyage round the World* (2000), 176.

104. Ibid., 150.

105. G. Forster to Campe, October 5, 1779, AA XIII, 243.

106. G. Forster, *Voyage round the World* (2000), 172.

107. Ibid., 81; Goethe, *Italian Journey*, 112, 188, 229.

108. Cf. Richter, *Goethe in Neapel*, 49.

109. G. Forster, *Voyage round the World* (2000), 51.

110. Ibid., 80.

111. Chamisso, *Voyage around the World*, 125.

112. For more on Wallace in Darwin's shadow, see Glaubrecht, *Am Ende des Archipels*.

113. Wallace, *Malay Archipelago*, 445.

114. Ibid.

115. Ibid.; Darwin, *Voyage of the Beagle*, 277.

116. G. Forster, "Der Brodbaum," AA VI/1, 69; G. Forster to C. G. Heyne, March 9, 1786, AA XIV, 446; G. Forster, *Voyage round the World* (2000), 43, 44.

117. G. Forster to T. Heyne, September 3, 1784, AA XIV, 180; G. Forster, *Ansichten vom Niederrhein*, AA IX, 29.

118. G. Forster, *Voyage round the World* (2000), 626; G. Forster, "Der Brodbaum," AA VI/1, 69; G. Forster, "Praelectiones zoologicae," AA VI/2, 1666. For this lecture, given in Wilna in 1786, Forster provided the second edition of Johann Friedrich Blumenbach's work *De generis hvmani varietate native liber* (Göttingen, 1781) as the template. It spoke of a "summa Dei prouidentia" (29). Also, Uhlig, "Hominis historia naturalis," 179. For a complex assessment of Forster's religiosity, cf. Dumont, "Seeing Is Believing." Dumont comes to the conclusion "that Forster increasingly found it difficult to accept the conceptions of God and the afterlife of his time, openly opposing 'offers of compromise' like theological rationalism" (196).

119. G. Forster, "Der Brodbaum," AA VI/1, 68.

120. G. Forster, "Reise von Kassel," *Tagebücher*, AA XII, 83; G. Forster to Spener, February 12, 1785, AA XIV, 279. Forster also speaks of the abundance of nature that cannot be understood by people and corrects us in our assumption of humanity's exceptional place. "We believe ourselves to be the ultimate purpose in the existence of all things surrounding us. A minimal degree of knowledge of nature can yank us out of this misconception. Everywhere we come across organizations, which we do not yet know, which we do not need to know, whose relationship to the rest of the earthly beings remains mysterious to us." G. Forster, *Ansichten vom Niederrhein*, AA IX, 64.

121. G. Forster, *Voyage round the World* (2000), 226.

122. Ibid., 121, 122.

123. Ibid., 200; G. Forster, "Preisverzeichnis von südlandischen Kunstsachen und Naturalien," AA V, 96.

124. G. Forster, "Neuguinea: Ein geographisches Fragment," AA V, 93.

125. Rousseau, *Discourse on Inequality*, 74.

126. Cf. Kronauer, "Rousseaus Kulturkritik." Kronauer suggests in his essay "Zurück zu den Affen" that Forster's voyage around the world "was not defined by Rousseauian optics" (91).

127. Cf. Kohl, *Entzauberter Blick*.

128. For Forster's "multi-perspectival depiction of foreigners as a program for inter-subjective understanding of their natures," see Strack, *Exotische Erfahrung und Intersubjektivität*, 233. Yomb May presents a literary examination of *Voyage round the World* with respect to Forster's incipient but inconsistent criticism of Eurocentrism in *Georg Forsters literarische Weltreise*.

129. G. Forster, *Voyage round the World* (2000), 25.

130. Ibid., 34.

131. Ibid., 34–35.

132. Ibid., 120–121, 123.

133. Ibid., 628.

134. Ibid., 630, 632.

135. Ibid., 144–145.

136. Ibid., 226.

137. Cf. Bödeker, "Die Natur des Menschen," 165.

138. Garber, "Reise nach Arkadien," 50.

139. Forster is sometimes thought to have accepted a tiered model without reservations. Cf. Holdenried, "Erfahrene Aufklärung," 141: "Like the Enlightenment, Forster generally proceeds from a societal hierarchy of individually visited peoples. It ranges in ascending order from the proverbial 'wretched' Pecherais, to the Marquesas Islanders to the idealized Tahitians." A judgment of this kind is evidenced by Forster's works. In his translator's introduction to George Keate's *Nachrichten von den Pelew-Inseln*, published in Hamburg in 1789, Forster writes of the respective "stages of development": "The New Zealanders and the Tanna Islanders were animal-like and savage; depraved and decadent in comparison to the higher cultures of the Tahitians and the inhabitants of the Friendly Islands." AA V, 327. Although Forster repeatedly made use of a tiered model, he did so inconsistently.

140. See Forster's reference in his letter to Jacobi of November 1, 1789, AA XV, 363; and his review from 1791, which critically engages with Meiners's publications in *Göttingischen historischen Magazin*, AA XI, 236–252. Cf. Sanches, "Dunkelheit und Aufklärung."

141. G. Forster, "Antwort an die Göttingischen Recensenten," AA IV, 53; G. Forster, "Der Brodbaum," AA VI/1, 64.

142. Garber, "Reise nach Arkadien," 50; G. Forster, *Voyage round the World* (2000), 210.

143. Garber, "Reise nach Arkadien," 50.

144. G. Forster, *Voyage round the World* (2000), 9.

145. Ibid., 631.

146. Ibid., 232.

147. Ibid., 240.

148. Schiller, "Cabal and Love," 372.

149. Kant, *Natural Science*, 504.

150. Zedler, *Grosses vollständiges Universal Lexicon*, vol. 20 (1995), "Menschen-Fresser" on 751; Polo, *Travels of Marco Polo*, 308.

151. G. Forster, *Voyage round the World* (2000), 103.

152. Ibid., 278, 283, 279.

153. Ibid., 278.

154. Ibid., 278–279. Forster's reaction to cannibalism reflects "the entire spectrum of European discourse on cannibalism," as Barbara Lichtblau shows in her dissertation, "Vorstellungen über Kannibalismus," 207.

155. G. Forster, *Voyage round the World* (2000), 279; Cf. Berg, *Zwischen den Welten*.

156. G. Forster, *Voyage round the World* (2000), 280, 278.

157. Ibid., 280–281; Montaigne, "On the Cannibals," 86–87.

158. G. Forster, *Voyage round the World* (2000), 280. During their stay on Tanna, an island in the New Hebrides in the South Pacific, "fifteen or twenty natives" threat-

ened that they would "be killed and eaten," should they venture any farther inland (524–525).

159. Ibid., 75.

160. Ibid., 605.

161. Ibid., 607.

162. G. Forster, *Reise um die Welt* (1989), 350.

163. G. Forster, *Voyage round the World* (2000), 606.

164. Ibid., 610.

165. G. Forster, "Über Leckereyen," AA VIII, 171; G. Forster, *Voyage round the World* (2000), 474.

166. Cf. Ritter, "Darstellungen der Gewalt."

167. G. Forster, *Voyage round the World* (2000), 39.

168. Ibid., 290.

169. Ibid., 148.

170. Ibid., 592, 253.

171. Ibid., 254.

172. Ibid., 253.

173. J. W. v. Goethe, *Aus meinem Leben: Dichtung und Wahrheit*, in Goethe, *Werke* (Weimar), 28:72; G. Forster, "Das Leben Dr. Wilhelm Dodds," AA VIII, 56, 59.

174. G. Forster, *Voyage round the World* (2000), 511.

175. The instructions are published in Cook, *Journals of Captain Cook*, 2:514–519. Force of arms is not to be used "till every other gentle method has been tried" (514).

176. G. Forster, *Voyage round the World* (2000), 407; Elliott, "Memoirs of John Elliott," 31: "Young Mr. Forster made his appearance, and fired, wounding the Man, who now retired, and joined his friends, with whom they had had some Skirmishing before. But had it not been for this providential circumstance, Cook's life would have been in most imminent danger."

177. G. Forster, *Voyage round the World* (2000), 549.

178. Ibid., 550. According to Ritter, in "Darstellungen der Gewalt," 37: "Presumably Forster has in mind an angel-pieta, a special variation of the *Imago pietatis* found in Italy, France, and Germany. In this depiction of the pieta, Christ's body is supported, carried, or displayed by one or more angels."

179. G. Forster, *Voyage round the World* (2000), 550.

180. Ibid., 550–551.

181. Reemtsma, "Mord am Strand," 41.

182. G. Forster, *Voyage round the World* (2000), 121.

183. G. Forster to J. R. Forster, December 14, 1778, AA XIII, 157; G. Forster to Zimmerman, March 13, 1788, AA XV, 127.

184. G. Forster, *Voyage round the World* (2000), 198–199.

185. Ibid., 162.

186. G. Forster, "Der Brodbaum," AA VI/1, 63, 79.

187. G. Forster, *Voyage round the World* (2000), 161, 198, 176.

188. Ibid., 199, 170, 177, 235.

189. Ibid., 270, 199.

190. Habermas, *Between Facts and Norms*, 132.

191. Rousseau, *Letter to D'Alembert*, 351.

192. Kleist, "Chilean Earthquake," 318.

193. G. Forster to Jacobi, November 2, 1779, AA XIII, 252; G. Forster, *Voyage round the World* (2000), 283.

194. G. Forster, *Voyage round the World* (2000), 548.

195. Ibid., 549; F. Schiller, "An die Freude" (1785), in *Sämtliche Werke*, 1:133.

196. G. Forster, *Voyage round the World* (2000), 199.

197. Ibid., 166.

198. Ibid., 164–165.

199. Ibid., 165. Cf. Forster's sketch of an anthropological cultural pessimism: "Thus, in every country, mankind are fond of being tyrants, and the poorest Indian, who knows no wants but those which his existence requires, has already learnt to enslave his weaker helpmate, in order to save himself the trouble of supplying those wants." *Voyage round the World* (2000), 583.

200. According to the reference by Uhlig in *Georg Forster*, 92.

201. G. Forster, *Cook, the Discoverer*, 245.

202. G. Forster, *Voyage round the World* (2000), 200.

III

1. For the whereabouts of the drawings, see Forster's letters to F. A. Vollpracht, September 1, 1776, AA XIII, 44; and to F. H. Jacobi, February 11, 1783, AA XIII, 428. Some of the drawings are printed in G. Forster, *Reise um die Welt: Illustriert von eigener Hand*. Of the more than forty duplicates and seven unique prints that were rediscovered in Sidney in 2007, some are reproduced in G. Forster, *Cook, der Entdecker*. See also *Forsters Bilder von der Weltumsegelung*.

2. G. Forster to Vollpracht, September 1, 1776, AA XIII, 44.

3. G. Forster to J. K. P. Spener, July 28, 1779, AA XIII, 229; December 15, 1779, AA XIII, 260 f., 262; July 5, 1779, AA XIII, 217.

4. G. Forster to C. G. Heyne, December 17, 1783, AA XIII, 513. For Forster's relationship to the Poles, see Fagot, "In den eigenen Grenzen exiliert."

5. G. Forster, "Reise von Kassel nach Wilna 1784," *Tagebücher*, AA XII, 180; G. Forster to Spener, December 17, 1778, AA XIII, 162.

6. Schiller, "Robbers," 70; G. Forster, "De plantis esculentis insularum oceani australis commentatio botanica," AA VI/1, 93–137; G. Forster to S. T. Soemmerring, February 3, 1785, AA XIV, 269.

7. G. Forster to Spener, September 12, 1777, AA XIII, 109.

8. G. Forster, "Reise von Kassel," *Tagebücher*, AA XII, 180.

9. G. Forster to Spener, June 5, 1779, AA XIII, 206.

10. G. Forster to J. R. Forster, September 8, 1783, AA XIII, 477; G. Forster to Spener, December 17, 1778, AA XIII, 162.

11. G. Forster to C. G. Heyne, October 12, 1786, AA XIV, 564.

12. G. Forster, "Reise von Kassel," *Tagebücher*, AA XII, 158.

13. Ibid., 95, 164, 59, 165. Recurring notes in journal entries from Forster's journey from Kassel to Vilna attest to his persistently depressive mood: "I cried too much yesterday" (AA XII, 21); "Sad" (AA XII, 87); "I was sad and somber inside all day" (AA XII, 90).

14. G. Forster to C. G. Heyne, May 1, 1792, AA XVII, 107.

15. G. Forster to T. Heyne, September 3, 1784, AA XIV, 179; December 13, 1784, AA XIV, 242; G. Forster to Soemmerring, February 3, 1785, AA XIV, 273; G. Forster to C. G. Heyne, March 9, 1786, AA XIV, 445; G. Forster to T. Heyne, December 13, 1784, AA XIV, 244; January 22, 1785, AA XIV, 264.

16. G. Forster to Soemmerring, February 3, 1785, AA XIV, 270.

17. G. Forster to Spener, April 10, 1786, AA XIV, 465.

18. G. Forster to C. G. Heyne, May 15, 1785, AA XIV, 319.

19. G. Forster, "Reise von Kassel," *Tagebücher*, AA XII, 116, 147.

20. G. Forster to Spener, March 5, 1788, AA XV, 117; September 17, 1776, AA XIII, 52; G. Forster to Vollpracht, May 10, 1776, AA XIII, 39; G. Forster, "Reise von Kassel," *Tagebücher*, AA XII, 137, 52, 63, 57, 71, 131; G. Forster to Soemmerring, August 20, 1788, AA XV, 184; G. Forster to J. Müller, December 20, 1783, AA XIII, 518; G. Forster, "Reise von Kassel," *Tagebücher*, AA XII, 62, 63.

21. G. Forster, "Reise von Kassel," *Tagebücher*, AA XII, 153.

22. G. Forster to Jacobi, November 16, 1782, AA XIII, 414.

23. G. Forster to Spener, October 22, 1776, AA XIII, 59.

24. *Johann Georg Forster's Briefwechsel: Nebst einigen Nachrichten fon seinem Leben* (Leipzig, 1829), 1:61, quoted in Steiner, *Georg Forster*, 96.

25. G. Forster, "Reise von Kassel," *Tagebücher*, AA XII, 67, 153. Ernst August Franke sums up that Forster, in the two decades following his voyage around the world, "was plagued by mentally and physically caused 'hypochondriac' disturbances from chronic gallbladder and joint disease, as well as related heart disease. Also the idea of a chronic tropical illness cannot be fully ruled out." Franke, "Georg Forsters Krankheiten und Tod," 220.

26. G. Forster to T. Heyne, January 22, 1785, AA XIV, 265; February 3, 1785, AA XIV, 274.

27. G. Forster, "Reise von Kassel," *Tagebücher*, AA XII, 148.

28. G. Forster to Soemmerring, July 18, 1785, AA XIV, 348.

29. Ibid., 349.

30. Ibid., 351.

31. According to Enzensberger, *Georg Forster*.

32. Reemtsma, "Mord am Strand," 52; G. Forster to Soemmerring, July 18, 1785, AA XIV, 349. In a letter to Soemmerring of November 20, 1786, he calculates that he still has eight to ten years to live (AA XIV, 590).

33. G. Forster, "Rundreise von Mainz aus 1790," *Tagebücher*, AA XII, 347; G. Forster, "Reise von Kassel," *Tagebücher*, AA XII, 22; G. Forster, "Rundreise von Mainz aus 1790," *Tagebücher*, AA XII, 266; G. Forster, "Reise von Kassel," *Tagebücher*,

AA XII, 52, 24; G. Forster, "Reise von Wilna nach Dresden 1785," *Tagebücher*, AA XII, 197.

34. G. Forster, "Reise von Kassel," *Tagebücher*, AA XII, 82.

35. G. Forster, *Ansichten vom Niederrhein*, AA IX, 12.

36. Ibid., 87.

37. Ibid., 53.

38. G. Forster, "Rundreise von Mainz aus 1790," *Tagebücher*, AA XII, 350.

39. G. Forster to T. Heyne, May 22, 1784, AA XIV, 70.

40. Cf. Steiner, *Freimaurer und Rosenkreuzer*; Reinalter, "Forster als Freimaurer und Rosenkreuzer."

41. Wieland, "Forsters Reise um die Welt," 143; G. Forster to C. G. Heyne, July 10, 1786, AA XIV, 508; G. Forster to C. G. Körner, November 25, 1789, AA XV, 376; G. Forster to C. G. Heyne, March 20, 1790, AA XVI, 33; G. Forster, preface to *Kleine Schriften: Ein Beitrag zur Völker- und Länderkunde, Naturgeschichte und Philosophie des Lebens*, AA V, 345. Wolf Lepenies also emphasizes Forster's anthropological interest as being central. Lepenies, "Forster als Anthropologe."

42. Pope, *Essay on Man*, 89.

43. Cf. Fink, "Klima und Kulturtheorien der Aufklärung," 25–55.

44. Bodin, *Six Books of the Commonwealth*, 145.

45. G. Forster, *Voyage round the World* (2000), 198.

46. Ibid., 142.

47. G. Forster, "Beobachtungen über das Klima von Senegal," AA V, 160, 143; Aristotle, *The Politics*, trans. Stephen Everson (Cambridge: Cambridge University Press, 1988); G. Forster, *Reise um die Welt*, ed. G. Steiner (Berlin: Akademie-Verlag, 1989), 272.

48. G. Forster, "Something More," 151.

49. G. Forster, *Voyage round the World* (2000), 592. Tanja van Hoorn reconstructs Forster's critical distance from climate theory during the voyage around the world in "Physische Anthropologie und normative Äesthetik."

50. Bodin, *Sechs Bücher über den Staat*, 2:159.

51. G. Forster, *Voyage round the World* (2000), 340.

52. G. Forster, "Versuch einer Naturgeschichte des Menschen: Anlagen in Menschen," AA VIII, 158.

53. G. Forster, *Voyage round the World* (2000), 151; G. Forster to Soemmerring, June 8–12, 1786, AA XIV, 486; G. Forster, description of the red tree creeper on the island O-Waihi, AA VI/1, 245; G. Forster, "Versuch einer Naturgeschichte des Menschen: Anlagen in Menschen," AA VIII, 159.

54. G. Forster, "Über lokale und allgemeine Bildung," AA VII, 45; G. Forster, *Voyage round the World* (2000), 467, 563.

55. Kant, "Determination of the Concept," 128.

56. Geulen, *Geschichte des Rassismus*, 48.

57. Cf. Godel and Stiening, *Klopffechtereien*.

58. Uhlig, *Georg Forster*, 202; Kühn, *Kant*, 398.

59. Kant, "Of the Different Human Races," 60, 62, 64–65, 60.

60. Kant, "Determination of the Concept," 128.

61. Ibid., 136; Kant, handwritten note, vol. 2/2, in Kant, *Gesammelte Schriften*, vol. 15/2:885. Cf. Forster's lecture of December 8, 1785, "De hominis in omni climate vivendi facultate," AA VI/2, 1047–1060.

62. Kant, "Use of Teleological Principles," 173.

63. G. Forster to Soemmerring, June 8–12, 1786, AA XIV, 486; G. Forster, *Cook, the Discoverer*, 199; Kant, "Determination of the Concept," 129; G. Forster, "Something More," 149; G. Forster to C. G. Heyne, August 10, 1786, AA XIV, 521; G. Forster to Soemmerring, January 19, 1787, AA XIV, 618; G. Forster to F. L. M. Meyer, August 10, 1786, AA XIV, 523.

64. Kant, "Determination of the Concept," 128; Kant, "Use of Teleological Principles," 174; G. Forster, "Something More," 148.

65. G. Forster, *Voyage round the World* (2000), 9.

66. G. Forster, "Something More," 166.

67. G. Forster, *Voyage round the World* (2000), 6.

68. G. Forster to Soemmerring, June 8–12, 1786, AA XIV, 486.

69. G. Forster, "Something More," 159.

70. Kant, "Of the Different Human Races," 71; G. Forster, "Something More," 156.

71. G. Forster, "Über lokale und allgemeine Bildung," AA VII, 51.

72. G. Forster to Jacobi, January 16, 1789, AA XV, 247.

73. Ibid., January 2, 1789, AA XV, 233.

74. G. Forster, "Something More," 155.

75. Home, *Sketches of the History of Man*, 3:142; Soemmerring to P. Camper, May 24, 1774, in Soemmerring, *Werke*, 18:479: "Aethiopum et Europaeum non varietate sed specie differre, et duos fuisse it ita dicam Adamos."

76. G. Forster to J. G. Herder, January 21, 1787, AA XIV, 621.

77. G. Forster, "Something More," 167, 162.

78. In his essay "Determination of the Concept of a Human Race," Kant was able to praise even more uncritically, with respect to the different human skin tones, the "wise arrangement of nature" and the "precaution of nature." Kant was aware at this time of not being able to prove teleologies of this kind, but of having to leave it to a "presumption of expediency." Kant, *Gesammelte Schriften*, 8:103, 93, 103.

79. Kant, *Critique of Judgment*, 184.

80. For example, Geulen, *Geschichte des Rassismus*, 52.

81. G. Forster, "Something More," 164, 165.

82. G. Forster, "Menschen-Racen," AA VIII 157. The note is undated; see G. Forster, "Erläuterungen," AA VIII, 404. It is nearly identical to a passage from a letter of Forster's to C. G. Heyne, November 20, 1786, AA XIV, 587.

83. Tanja van Hoorn points to this important fact in "Dem Leibe abgelesen," 124–128.

84. G. Forster to J. G. Herder, January 21, 1787, AA XIV, 621.

85. G. Forster to J. H. Merck, September 10, 1786, AA XIV, 543.

86. G. Forster to Soemmerring, July 23, 1786, AA XIV, 515.

87. G. Forster to S. v. La Roche, March 19, 1790, AA XVI, 32. In this letter he discusses his essay "Die Kunst und das Zeitalter," AA VII, 15–26.

88. G. Forster to Soemmerring, July 23, 1786, AA XIV, 515.

89. G. Forster to Spener, November 13, 1786, AA XIV, 581.

90. G. Forster to C. G. Heyne, November 20, 1786, AA XIV, 587.

91. G. Forster to Jacobi, November 19, 1788, AA XV, 208; Grimm, *Deutsches Wörterbuch*, vol. 5, *Klopffechter*, 1229 f.

92. G. Forster to Jacobi, November 19, 1788, AA XV, 208; February 8, 1789, AA XV, 264.

93. G. Forster to Soemmerring, January 19, 1787, AA XIV, 618. For the significant relationship between Forster and Herder, see Henning, "'Vortrefflicher Mann' und 'bester Freund'"; Uhlig, "Georg Forster und Herder"; May, "'Ganz ist Herder doch mein Mann nicht.'"

94. In addition, Enke, "'Praelectiones anatomicae.'" Soemmerring was Forster's closest friend; F. Dumont offers a characterization of this relationship and a portrait of Soemmerring, who mostly stood in Forster's shadow, in "Das 'Seelenbündnis.'"

95. S. T. Soemmerring, "Ueber die körperliche Verschiedenheit des Negers vom Europäer," in Soemmerring, *Werke*, 15:151; Kant, "Determination of the Concept," 139.

96. Kant, "Of the Different Human Races," 67.

97. Ibid., 69:

The region of the earth between 31 and 52 degrees latitude in the Old World (which also seems to deserve the name Old World with regard to the population [living there]) can, however, be thought of as one in which the most fortunate mixture of the influences of [both] the colder and the hotter regions and also the greatest riches in earthly creatures are to be found. [This is] also [the region] where human beings would have to diverge least from their original formation because [the human beings living] in this region are equally well-prepared for any transplantation from there outward.

98. G. Forster to T. Heyne, January 24, 1785, AA XIV, 266.

99. G. Forster, "Something More," 165.

100. Soemmerring, "Ueber die körperliche Verschiedenheit," 15:247.

101. So I assumed when first approaching Forster's thought; I was ignoring his consideration of polygenesis and defense of monogenesis in the context of slavery: Goldstein, *Die Entdeckung der Natur*, 83.

102. G. Forster, "Über Leckereyen," AA VIII, 167 f.

103. G. Forster, "Praelectiones zoologicae," AA VI/2, 1693–1701. The seven *Exempla hominum* were (1) Greenlanders, (2) Americans, (3) Scythians, Mongols, Kalmyks, and Chinese, (4) Indians, (5) Caucasians, (6) Africans, (7) Australians.

104. G. Forster, "Rudimenta zoologicae," AA VI/2, 1507; "Praelectiones zoologicae," AA VI/2, 1670–1673.

105. G. Forster, "Praelectiones zoologicae," AA VI/2, 1692: "Hinc nihil nobis praeter conjecturas relictum est, si de primitivis varietatibus disceptare volumus; sin secus

nulla hypothesium ratione habita, ea quae nunc innotescunt *Exempla*, quotquot sunt insigniora, enumerare debemus juxta evidentiorem formae diversitatem et singulares atque characteristicas notas."

106. This is presented in immaculate detail by Uhlig, "Hominis historia naturalis." For an analysis of Forster's lectures on natural history since his time in Wilna, see Jahn, "'Scienta Naturae.'"

107. Soemmerring, "Ueber die körperliche Verschiedenheit," 15:162. In his lecture "Praelectiones zoologicae" (AA VI/I, 1673), Forster considers the question of whether there are one or more than one species of human beings. According to his response, there can be found throughout the whole world people with the same number, shape, and proportion of body parts, such that all can be classified into one species: "Atque adeo omnes ad unam speciem referendos esse."

108. G. Forster, *Voyage round the World* (2000), 374.

109. G. Forster to J. B. Jachmann, between September 12 and October 14, 1790, AA XVI, 185.

110. G. Forster to Spener, March 12, 1788, AA XV, 124; G. Forster to Jacobi, November 19, 1788, AA XV, 210. In a letter to his father-in-law, Christian Gottlob Heyne, October 24, 1791, Forster sums up the influence of his psycho-physical temperament on his thinking: "An infirm disposition alone wrecks philosophy" (AA XVI, 360).

111. I. Kant, *Physische Geographie*, in Kant, *Gesammelte Schriften*, 9:198. Wolfdietrich Schmied-Kowarzik seeks to reconcile the closeness of Forster's, Kant's, and Herder's positions in "Der Streit um die Einheit."

112. Yomb May attempts to trace Forster's thought process, "in which he realizes the diversity of the human race and the requirement for its recognition." May, "Pluralität und Ethos," 354.

113. G. Forster to C. G. Heyne, August 30, 1790, AA XVI, 176; also, the preceding letter from Heyne to Forster, August 16, 1790, AA XVIII, 415–417, in which Heyne speaks of Kant.

114. I. Kant, "Idee zu einer allgemeinen Geschichte in weltbürgerlicher Absicht," in Kant, *Gesammelte Schriften*, 8:29; I. Kant, "Review of Herder's *Ideas for a Philosophy of the History of Mankind*," in Kant, *On History*, 50–51.

115. G. Forster, "Darstellung der Revolution in Mainz," AA X/1, 513.

116. Ibid., 514; Kant, "Review of Herder's *Ideas*," 27.

117. G. Forster, "Über lokale und allgemeine Bildung," AA VII, 49.

118. G. Forster, "Something More," 150.

119. G. Forster to Spener, September 1, 1779, AA XIII, 232.

120. Gilbert, *Captain Cook's Final Voyage*, 107; G. Forster, "Fragmente über Capitain Cooks letzte Reise und sein Ende," AA V, 72–92.

121. G. Forster to Spener, December 4, 1786, AA XIV, 595; see G. Forster to Spener, April 1, 1787, AA XIV, 659.

122. G. Forster to Jacobi, January 2, 1789, AA XV, 232; September 18, 1790, AA XVI, 191. In a letter to Meyer of April 2, 1787 (AA XIV, 662), Forster writes the contrary, that it is not a Panegyric to Cook. The apparent contradiction can be taken as proof of the two levels on which *Cook, the Discoverer* operates.

123. G. Forster to Soemmerring, April 29, 1786, AA XIV, 472; G. Forster to Spener, April 8, 1787, AA XIV, 669.

124. G. Forster, *Cook, the Discoverer*, 184, 187. See Reemtsma, "Mord am Strand," 55: "Georg attached himself to Cook as far as possible; he performed something like an adoption in spirit and offered himself as the son who can take better care of his legacy than anyone else."

125. G. Forster to Meyer, April 2, 1787, AA XIV, 662; G. Forster, *Cook, the Discoverer*, 243.

126. G. Forster, *Cook, the Discoverer*, 170, 197, 155.

127. Ibid., 162.

128. Ibid., 259.

129. In a letter to Meyer of April 2, 1787 (AA XIV, 663), Forster discloses this motive by explicitly naming Kant: "And you will probably notice how I campaigned against him in my essay on Cook."

130. Garber, "Reise nach Arkadien," 32.

131. G. Forster, *Voyage round the World* (2000), 481.

132. Ibid., 456; G. Forster, *Cook, the Discoverer*, 244.

133. G. Forster, *Cook, the Discoverer*, 225.

134. Ibid., 228.

135. G. Forster, *Voyage round the World* (2000), 481.

136. G. Forster, "New Holland and the British Colony at Botany Bay," trans. Robert J. King, posted on April 18, 2008, www.australiaonthemap.org.au/new-holland -and-the-british-colony-at-botany-bay-2/; Ferguson, *Essay on Civil Society*, 12.

137. G. Forster, "New Holland and Botany Bay."

138. G. Forster, *Cook, the Discoverer*, 197.

139. Forster placed the motto *Nullius in Verba* at the front of the commemorative volume, meaning to swear on no one's word, to let experience alone speak.

140. G. Forster, *Cook, the Discoverer*, 224, 228, 212.

141. Ibid., 224.

142. Ibid., 212.

143. Ibid., 246.

144. G. Forster, "Über die Beziehung der Staatskunst auf das Glück der Menschheit," AA X/1, 591.

145. G. Forster, *Cook, the Discoverer*, 158.

146. Ibid., 251.

147. Ibid., 213; G. Forster, "Über Leckereyen," AA VIII, 174.

148. G. Forster, *Cook, the Discoverer*, 253.

149. Ibid., 246.

150. Ibid., 221, 216.

151. Ibid., 264.

152. Ibid., 259.

153. G. Forster, "Geschichte der Englischen Litteratur, vom Jahr 1788," AA VII, 62.

154. G. Forster, *Cook, the Discoverer*, 259.

155. G. Forster to Jacobi, December 20, 1783, AA XIII, 514 f.; G. Forster, *Cook, the Discoverer*, 252; G. Forster to J. R. Forster, February 13, 1783, AA XIII, 430.

156. G. Forster to J. R. Forster, February 13, 1783, AA XIII, 430.

157. G. Forster, *Cook, the Discoverer*, 257.

158. G. Forster to C. G. Heyne, September 1, 1784, AA XIV, 178.

IV

1. G. Forster to C. G. Heyne, July 30, 1789, AA XV, 319.

2. C. G. Heyne to G. Forster, August 2, 1789, AA XVIII, 341.

3. A. v. Humboldt to G. Forster, August 7, 1789, AA XVIII, 341.

4. G. Forster to C. W. v. Dohm, July 13–18, 1790, AA XVI, 158; G. Forster to F. H. Jacobi, January 2, 1789, AA XV, 231; G. Forster to C. F. Voss, November 10, 1792, AA XVII, 242; G. Forster to Jacobi, January 2, 1789, AA XV, 231; G. Forster to T. Forster, February 4, 1793, AA XVII, 326; G. Forster to Voss, November 21, 1792, AA XVII, 252; G. Forster to T. Forster, January 2, 1793, AA XVII, 296; G. Forster to C. G. Heyne, November 10, 1792, AA XVII, 238.

5. G. Forster, "Über historische Glaubwürdigkeit," AA VII, 30; "Can the world ever reasonably and through reason become happy?," AA VIII, 358.

6. G. Forster, "Parisische Umrisse," AA X/1, 602. Forster had previously spoken of a "public opinion," for example, in the letter to Christian Gottlob Heyne of June 5, 1792, AA XVII, 128; cf. also the letter fragment, probably after 1790: "Über die öffentliche Meinung," AA VIII, 364; Habermas, *Structural Transformation of Public Sphere*, 101; G. Forster, "Parisische Umrisse," AA X/1, 602.

7. G. Forster, *Ansichten vom Niederrhein*, AA IX, 111, 108.

8. Ibid., 111.

9. *Deutsche Chronik*, November 2, 1775; March 25, 1776; November 2, 1775, all quoted in Warneken, *Schubart*, 137, 155, 137; Habermas, *Structural Transformation of Public Sphere*, 73: "Even the brutal reaction of the princes against the first political publicists in southwestern Germany was symptomatic of a certain critical strength of the public sphere."

10. *Die neue Mainzer Zeitung oder der Volksfreund*, no. 1, January 1, 1793, AA X/1, 173; G. Forster, "Rundreise von Mainz aus 1790," *Tagebücher*, AA XII, 293; *Die neue Mainzer Zeitung oder der Volksfreund*, no. 1, January 1, 1793, AA X/1, 186. The number of new literary and political periodicals speaks to the newly forming public in the context of the French Revolution; cf. Hocks and Schmidt, *Literarische und politische Zeitschriften*.

11. According to Thamer, *Die Französische Revolution*, 17.

12. Ibid., 20; Soboul, *Die Große Französische Revolution*, 74, 77, 82.

13. Sièyes, *What Is the Third Estate?*, 53, 57.

14. Cf. Revel, "Die Große Angst," 1:110–121; Ikni, "Grande Peur," 517.

15. Soboul, *Die Große Französische Revolution*, 125.

16. G. Forster to Jacobi, September 21, 1789, AA XV, 338.

17. Ibid., December 8, 1789, AA XV, 380.

18. G. Forster to Voss, November 10, 1792, AA XVII, 240.

19. G. Forster to Jacobi, March 9, 1784, AA XIV, 32; Kant, *Critique of Pure Reason*, 478.

20. Kant, *Religion within the Boundaries*, 68.

21. Koselleck, *Futures Past*, 64.

22. Jaucourt, "Revolution (Moderne Geschichte Englands)," 360.

23. G. Forster to C. G. Heyne, June 5, 1792, AA XVII, 128.

24. G. Forster to Jacobi, January 2, 1789, AA XV, 231; G. Forster to C. G. Heyne, September 1, 1792, AA XVII, 168.

25. Kant, *Conflict of the Faculties*, 151.

26. G. Forster to C. G. Heyne, July 12, 1791, AA XVI, 313.

27. G. Forster, *Voyage round the World* (2000), 200.

28. G. Forster to C. G. Heyne, July 30, 1789, AA XV, 319.

29. G. Forster, *Voyage round the World* (2000), 200; G. Forster, *Cook, the Discoverer*, 253.

30. G. Forster to Jacobi, January 2, 1789, AA XV, 231.

31. G. Forster to T. Forster, April 14, 1788, AA XV, 147.

32. G. Forster to C. G. Heyne, January 9, 1789, AA XV, 240.

33. G. Forster to Jacobi, November 15, 1789, AA XV, 371; G. Forster, *Ansichten vom Niederrhein*, AA IX, 291; G. Forster to S. v. La Roche, March 19, 1790, AA XVI, 32.

34. G. Forster to S. v. La Roche, March 19, 1790, AA XVI, 32. For the professional and personal closeness of Forster and Humboldt, cf. Beck, "Georg Forster und Alexander von Humboldt"; Graczyk, "Forschungsreise und Naturbild."

35. G. Forster to Jacobi, April 6, 1792, AA XVII, 91; G. Forster to Voss, March 6, 1792, AA XVII, 58. For the stylistic focus of *Ansichten vom Niederrhein* on its audience, cf. Wuthenow, "Enzyklopädische Reisebeschreibung." Fischer gives an overall interpretation of *Ansichten* in *Reisen als Erfahrungskunst*.

36. G. Forster to S. T. Soemmerring, October 10, 1784, AA XIV, 189; G. Forster to J. K. P. Spener, July 23, 1790, AA XVI, 164; G. Forster to Voss, November 27, 1790, AA XVI, 208. Lichtenberg predicted success for this calculation: "The superlative printing will move some to purchase it, which will not affect the immortality of the work itself and cannot diminish it." G. C. Lichtenberg to Soemmerring, April 20, 1791, in Soemmerring, *Werke*, vol. 19/2, p. 839.

37. *Deutsche Chronik*, July 12, 1791; see G. Steiner's remarks on the history of its impact in his introduction to Forster's *Anischten vom Niederrhein*, AA IX, 356; Lichtenberg to G. Forster, July 1, 1791, AA XVIII, 452; J. W. v. Goethe to G. Forster, June 25, 1792, AA XVIII, 540.

38. C. G. Heyne to G. Forster, August 22, 1790, AA XVIII, 417, 418; April 12, 1792, AA XVIII, 516.

39. G. Forster to Spener, July 23, 1790, AA XVI, 162. This does not deny that for Forster art possessed great importance, not least for the process of humanization in light of a one-sided rationalization. Cf. Hoorn, "Zwischen Humanitätsideal und Kunstkritik"; Fischer, "'Wer ist der hohe Fremdling"; Gilli, "Georg Forster als Kunstkritiker."

40. G. Forster to Spener, July 23, 1790, AA XVI, 162; G. Forster, "Rundreise von Mainz aus 1790," *Tagebücher*, AA XII, 226.

41. G. Forster to C. G. Heyne, April 21, 1792, AA XVII, 101.

42. Cf. G. Steiner's references with regard to this in his introduction to Forster's *Ansichten vom Niederrhein* (AA IX, 346), in *Georg Forster*, 63; Koch, "Selbstbildung und Leserbildung," 24–27.

43. Steiner, *Georg Forster*, 63.

44. G. Forster to F. Schiller, December 7, 1790, AA XVI, 212; G. Forster to Jacobi, April 6, 1792, AA XVII, 91; G. Forster to Voss, December 18, 1790, AA XVI, 216; Enzensberger, *Georg Forster*, 207; The dedication reads: "In the wanderer's breast you make a more beautiful law of sensations come true. Their creation is dedicated to you! Let their blessing increase in value so that something of the gift may be inherent in the giver. Is the priest merely being bold when he empties the offering bowl to his guardian angel in front of his congregation? Or does anyone sense in one lucrative glance the great, pure, hushed pleasure of his completion?" AA IX, iv. See Forster's explanation to Voss in AA XVI 218. The dedication is intended for Forster's wife, Therese, who— pregnant once again—had openly begun a love affair with Ludwig Ferdinand Huber in Forster's absence while he was on his Rhine tour.

45. G. Forster, "Rundreise von Mainz aus 1790," *Tagebücher*, AA XII, 226.

46. G. Forster, *Cook, the Discoverer*, 259; G. Forster to Jacobi, March 22, 1791, AA XVI, 252. Marita Gilli works out to what extent the many reviews that Forster wrote can be viewed not just as supplementary work but as a part of his work on political publics—at least insofar as they do not concern purely geographical new publications; cf. Gilli, "Das Politische in Forsters Rezensionen."

47. G. Forster to C. G. Heyne, March 17, 1792, AA XVII, 75; C. G. Heyne to G. Forster, April 12, 1792, AA XVIII, 516; G. Forster, *Ansichten vom Niederrhein*, AA IX, 317, 104.

48. G. Forster, *Ansichten vom Niederrhein*, AA IX, 7.

49. Rousseau, "Social Contract," 156; Lichtenberg to Soemmerring, April 20, 1791, in Soemmerring, *Werke*, vol. 19/2, 839.

50. G. Forster, *Ansichten vom Niederrhein*, AA IX, 7.

51. *Deutsche Chronik*, July 12, 1791, quoted from G. Steiner's remarks about the history of its influence in his introduction to Forster's *Ansichten vom Niederrhein*, AA IX, 357; G. Forster, *Ansichten vom Niederrhein*, AA IX, 104.

52. G. Forster, *Ansichten vom Niederrhein*, AA IX, 104.

53. Ibid.; G. Forster to T. Forster, April 1, 1790, AA XVI, 50.

54. G. Forster to T. Forster, April 1, 1790, AA XVI, 50; G. Forster, *Ansichten vom Niederrhein*, AA IX, 104.

55. G. Forster, *Ansichten vom Niederrhein*, AA IX, 23.

56. Ibid., 24.

57. Ibid., 36.

58. Ibid., 37. Forster repeatedly quoted from Goethe's *Ode to Prometheus* in his work; cf. Peitsch, "Forster und Goethes 'Prometheus.'"

59. G. Forster, *Ansichten vom Niederrhein*, AA IX, 149; G. Forster to J. R. Forster,

March 22, 1784, AA XIV, 35; G. Forster to Soemmerring, February 3, 1785, AA XIV, 273.

60. G. Forster, *Ansichten vom Niederrhein*, AA IX, 109.

61. Ibid., 96.

62. Ibid., 99.

63. Ibid., 299.

64. Forster describes Herschel's telescope in his journal "Rundreise von Mainz aus 1790," AA XII, 312–315.

65. G. Forster to Voss, September 4, 1790, AA XVI, 183.

66. According to Segeberg's apt formulation in "Georg Forsters Ansichten vom Niederrhein," 15.

67. Schlegel, "Georg Forster," part 1, 2:91.

68. Ibid. Cf. Pickerodt, "Der 'gesellschaftliche Schriftsteller.'"

69. In this sense it falls short, as Gerhart Pickerodt understands Forster's writing to be an expression of political authorship only starting with the occupation of Mainz by the French on October 21, 1792, when *Anischten vom Niederrhein* was already a political work; cf. Pickerodt, "Forster als politischer Schriftsteller." Cf. also Rasmussen, "Forster als gesellschaftlicher Schriftsteller."

70. Schlegel, "Georg Forster," 98; G. Forster, *Ansichten vom Niederrhein*, AA IX, 87.

71. Schlegel, "Georg Forster," 87.

72. G. Forster, *Ansichten vom Niederrhein*, AA IX, 294, 129.

73. Ibid., 112, 161.

74. Ibid., 128; G. Forster, *Voyage round the World* (2000), 200; G. Forster, *Ansichten vom Niederrhein*, AA IX, 319, 129, 161.

75. G. Forster to Dohm, mid-July, 1790, AA XVI, 159.

76. G. Forster, *Ansichten vom Niederrhein*, AA IX, 85; F. Schiller, "Über die ästhetische Erziehung des Menschen in einer Reihe von Briefen," in *Sämtliche Werke*, 5:575; Marita Gilli interprets *Ansichten vom Niederrhein* in light of Forster's hesitant thinking as a contradictory work that oscillates between reform and revolution: Gilli, "Reform und Revolution bei Forster."

77. G. Forster, *Ansichten vom Niederrhein*, AA IX, 113.

78. Ibid., 118.

79. Ibid.

80. Ibid., 12, 13; G. Forster, *Voyage round the World* (2000), 510, 517, 524, 525.

81. G. Forster, *Ansichten vom Niederrhein*, AA IX, 318; Cf. Fritscher, "Die Entmoralisierung der Naturgewalten"; Fritscher, "Vulkanismusstreit und Französische Revolution."

82. G. Forster, *Ansichten vom Niederrhein*, AA IX, 129.

83. G. Forster to Jacobi, January 2, 1789, AA XV, 231.

84. G. Forster, *Cook, the Discoverer*, 256.

85. G. Forster, "Über lokale und allgemeine Bildung," AA VII, 48.

86. G. Forster, "Die Nordwestküste von Amerika, und der dortige Pelzhandel,"

AAV, 427. This text deals with a section of Forster's compilation *Geschichte der Reisen, die seit Cook and der Nordwest- und Nordost-Küste von Amerika und in dem nördlichsten Amerika selbst von Meares, Dixon, Portlock, Cose, Long u.a.m. unternommen worden sind,* published in Berlin in 1791.

87. G. Forster, "Die Nordwestküste von Amerika," AAV, 512; G. Forster, "Leitfaden zu einer Geschichte der Menschheit," AA VIII, 190.

88. G. Forster, "Versuch einer Naturgeschichte des Menschen: Anlagen im Menschen," AA VIII, 158.

89. G. Forster, "Erinnerungen aus dem Jahr 1790 in historischen Gemälden und Bildnissen von D. Chodowiecki, D. Berger, C. Kohl, J. F. Bolt, und J. S. Ringck," AA VIII, 267; G. Forster to C. G. Heyne, July 3, 1792, AA XVII, 139; November 10, 1792, AA XVII, 237.

90. G. Forster to Jacobi, October 4, 1790, AA XVI, 193; G. Forster to C. G. Heyne, July 3, 1792, AA XVII, 139; June 5, 1792, AA XVII, 128; Oliver Hochadel's extremely lucid account, "Natur—Vorsehung—Schicksal," 90.

91. G. Forster to C. G. Heyne, February 21, 1792, AA XVII, 49; G. Forster to Dohm, April 5, 1791, AA XVI, 265; G. Forster to M. W. Thun, October 12, 1784, AA XIV, 196.

92. G. Forster, *Voyage round the World* (2000), 381; G. Forster, *Reise um die Welt,* ed. G. Steiner (Berlin: Akademie-Verlag, 1989), 378; G. Forster, "Something More," 167.

93. G. Forster, *Cook, the Discoverer,* 155.

94. G. Forster to C. G. Heyne, July 12, 1791, AA XVI, 314.

95. Ibid., January 7, 1792, AA XVII, 26.

96. Cf. Hochadel, "Natur—Vorsehung—Schicksal," 90: The term "fate" becomes synonymous with nature.

97. G. Forster to Jacobi, August 29, 1783, AA XIII, 472.

98. Ibid., 473.

99. G. Forster, *Cook, the Discoverer,* 155, 239, 240.

100. G. Forster to T. Forster, June 10, 1793, AA XVII, 365.

101. G. Forster, "Ein Blick in das Ganze der Natur: Einleitung zu Anfangsgründen der Thiergeschichte," AA VIII, 78, 81; G. Forster to C. G. Heyne, July 3, 1792, AA XVII, 138.

102. G. Forster, "Geschichte der Englischen Litteratur vom Jahre 1790," AA VII, 191.

103. G. Forster to T. Forster, June 10, 1793, AA XVII, 364; September 6, 1793, AA XVII, 439; G. Forster to T. Forster and L. F. Huber, December 28, 1794, AA XVII, 498.

104. Goethe to W. v. Humboldt, October 19, 1830, in "Goethe's Correspondence with Humboldt," 484.

105. Goethe to K. F. Zelter, October 5, 1830, in Goethe, "Goethe's Correspondence with Zelter." For the political metaphor of the earthquake, cf. Demandt, *Metaphern für Geschichte,* 137.

106. Marx, *Eighteenth Brumaire of Louis Bonaparte,* 8.

107. G. Forster, "Historisches Jahrgemälde von 1790: I, Revolutionen und Gegen-

revolutionen," AA VIII, 234; G. Forster, *Ansichten vom Niederrhein*, AA IX, 124; G. Forster to L. F. Huber, November 5, 1792, AA XVII, 234; G. Forster to C. G. Heyne, November 10, 1792, AA XVII, 236; G. Forster, *Ansichten vom Niederrhein*, AA IX, 86.

108. G. Forster, "Rundreise von Mainz aus 1790," *Tagebücher*, AA XII, 264.

109. G. Forster, "Anrede an die Gesellschaft der Freunde der Freiheit und Gleichheit am Neujahrstage 1793," AA X/1, 61.

110. G. Forster to J. F. Mieg, January 11, 1793, AA XVII, 307; G. Forster to C. G. Heyne, November 10, 1792, AA XVII, 237; G. Forster, "Über die Beziehung der Staatskunst auf das Glück der Menschheit," AA X/1, 587, 571.

111. For Herder's organic metaphorics, cf. the comprehensive account and numerous examples in V. Albus's study *Weltbild und Metapher*, 288–324; for the metaphor of fermentation, 310–315.

112. Zedler, "Fermentation, Gährung," in *Grosses vollständiges Universal Lexicon* (1994), 9:578; G. Forster, *Ansichten vom Niederrhein*, AA IX, 328.

113. Zedler, "Fermentation, Gährung," 9:578; G. Forster, "Über die Vernunft, in Beziehung auf das Glück der Menschheit," AA VIII, 362; G. Forster, *Ansichten vom Niederrhein*, AA IX, 129, AA IX, 123.

114. G. Forster, *Ansichten vom Niederrhein*, AA IX, 123; the memoirs of the cardinal of Retz were published in 1717 in Amsterdam.

115. G. Forster, "Rundreise von Mainz aus 1790," *Tagebücher*, AA XII, 256.

116. G. Forster to Jacobi, November 23 [?], 1789, AA XV, 374; G. Forster, "Historisches Jahrgemälde von 1790," AA VIII, 235.

117. G. Forster, "Parisische Umrisse," AA X/1, 607.

118. For organic and mechanical metaphorics, cf. Blumenberg, *Paradigms for a Metaphorology*, 62–76; also instructive is the study by Meyer, "Metaphorik politischer Philosophie." See also Garber, "Anthropologie und Geschichte."

119. Aristotle, *The Politics*, ed. Stephen Everson (Cambridge: Cambridge University Press, 1988); Blumenberg, *Paradigms for a Metaphorology*, 64.

120. Hobbes, *On the Citizen*, 10.

121. G. Forster, "Anrede an die Gesellschaft," AA X/1, 61.

122. G. Forster, "Über lokale und allgemeine Bildung," AA VII, 51.

123. G. Forster, "Erinnerungen aus dem Jahr 1790," AA VIII, 327.

124. G. Forster to T. Forster, April 11, 1788, AA XV, 142.

125. G. Forster to J. G. Herder, December 10, 1791, AA XVI, 392; G. Forster to Schiller, December 7, 1790, AA XVI, 212.

126. G. Forster, "Unterthänigste Pro Memoria," September 9, 1792, AA XVII, 181. Cf. Mathy, "Die letzten Aktivitäten Georg Forsters."

127. For Goethe's handling of Schiller's skull and the difficulties of clearly identifying it, cf. Schöne, *Schillers Schädel*.

128. Goethe to Grand Duke Carl August, September 27, 1826, in Goethe, *Werke* (Weimar), part 4, 41:177; G. Forster, "Unterthänigste Pro Memoria," September 9, 1792, AA XVII, 181.

129. The anecdote is reported by Uhlig, *Georg Forster*, 124; G. Forster to Dohm, April 5, 1791, AA XVI, 267.

130. G. Forster to Jacobi, November 6, 1791, AA XVI, 367; G. Forster to Friedrich Wilhelm II of Prussia, AA XVII, 41.

131. G. Forster to Jacobi, November 6, 1791, AA XVI, 366.

132. G. Forster to Dohm, April 5, 1791, AA XVI, 267; G. Forster to Jacobi, August 9, 1791, AA XVI, 328.

133. G. Forster to C. G. Heyne, April 10, 1792, AA XVII, 93.

134. G. Forster to Voss, April 17, 1792, AA XVII, 100; G. Forster to C. G. Heyne, April 21, 1792, AA XVII, 102.

135. G. Forster to C. G. Heyne, April 28, 1792, AA XVII, 106.

136. For historical details and an overview of the history of the Mainz Republic: Dumont, *Die Mainzer Republik*; for the siege of Mainz specifically, 58.

137. G. Forster to C. G. Heyne, October 5, 1792, AA XVII, 101; G. Forster to Voss, October 21, 1792, AA XVII, 207, 209; November 21, 1792, AA XVII, 250; G. Forster, "Anrede an die Gesellschaft," AA X/1, 61.

138. G. Forster to C. G. Heyne, February 21, 1792, AA XVII, 46; G. A. Craig, "A German Jacobin: Georg Forster," in Craig, *Politics of the Unpolitical*, 31; G. Forster to J. v. Müller, September 10, 1792, AA XVII, 175.

139. Dumont, *Die Mainzer Republik*, 58.

140. G. Forster to Voss, October 21, 1792, AA XVII, 211; October 27, 1792, AA XVII, 225; G. Forster to C. G. Heyne, October 22, 1792, AA XVII, 213; G. Forster to Voss, October 27, 1792, AA XVII, 226; Craig, "German Jacobin," 32.

141. Dumont, *Die Mainzer Republik*, 133.

142. Ibid., 135. According to Dumont, the Enlightenment is the departure point for Forster's political development. See Dumont, "Georg Forster als Demokrat."

143. G. Forster to C. G. Heyne, July 30, 1789, AA XV, 319; G. Forster, "Erinnerungen aus dem Jahr 1790," AA VIII, 267; G. Forster to C. G. Heyne, November 10, 1792, AA XVII, 237.

144. G. Forster to Lichtenberg, May 10, 1792, AA XVII, 109; G. Forster to T. Forster, September 25, 1793, AA XVII, 447; G. Forster to Voss, October 21, 1792, AA XVII, 210.

145. G. Forster to Voss, October 27, 1792, AA XVII, 226.

146. G. Forster, "Über das Verhältnis der Mainzer gegen die Franken," AA X/1, 13, 15, 25, 13.

147. As Forster describes it in a letter to Huber, October 24, 1792, AA XVII, 214.

148. G. Forster, "Über das Verhältnis der Mainzer," AA X/1, 16.

149. G. Forster to Huber, December 4, 1792, AA XVII, 257; G. Forster to Soemmerring, January 6, 1793, AA XVII, 300.

150. G. Forster to Huber, November 5, 1792, AA XVII, 235.

151. G. Forster to J. R. Forster, November 26, 1792, AA XVII, 254; G. Forster to Voss, November 10, 1792, AA XVII, 242; G. Forster, "Anrede an die Gesellschaft," AA X/1, 60. Forster had already spoken out against the use of corsets by women in his essay "Über die Schädlichkeit der Schnürbrüste," since the "powers of imagination are fastened by the small hook on the shoulder blade" (AA VIII, 183), thereby inhibiting

mobility. This rejection of the corset already has undertones of free thinking and thus political mobility. Cf. Gilli, "Auf dem Weg von der Wissenschaft."

152. G. Forster to A. W. Iffland, July 13–30, 1790, AA XVI, 160; Iffland to G. Forster, July 30, 1790, AA XVIII, 412; G. Lichtenberg to G. Forster, September 30, 1790, AA XVIII, 426; G. Forster, "Über Proselytenmacherei," AA VIII, 202.

153. G. Forster to Müller, September 10, 1792, AA XVII, 175.

154. G. Forster, "Über Proselytenmacherei," AA VIII, 202.

155. G. Forster to Huber, December 28, 1792, AA XVII, 288; Dumont, *Die Mainzer Republik*, 195; G. Forster to Huber, December 28, 1792, AA XVII, 288; Dumont, *Die Mainzer Republik*, 203.

156. G. Forster to T. Forster, December 8, 1792, AA XVII, 264; G. Forster to Voss, December 21, 1792, AA XVII, 279; G. Forster to T. Forster, December 8, 1792, AA XVII, 264. Forster fluctuates between making use of and rejecting the despotism of liberty, when he explains in a letter to Joseph Anton Dorsch, January 17, 1793: "I am not for, I have never been for, nor will I ever be for coercing public opinion; everything takes place voluntarily." AA XVII, 308.

157. Craig, "German Jacobin," 26; Schiller to C. G. Körner, December 21, 1792, in Schiller, *Werke*, 26:171.

158. G. Forster to Voss, November 10, 1792, AA XVII, 239. Forster can just as well claim the opposite. A few months later, on July 7, 1793, he wrote in a letter to his wife, Therese: "My unhappiness is the work of my principles, not my passions. I cannot act any different, and begin again." AA XVII, 382.

159. G. Forster, *Cook, the Discoverer*, 160.

160. G. Forster to Voss, June 4, 1791, AA XVI, 298; Paine, *Rights of Man*, 26; G. Forster to Voss, June 4, 1791, AA XVI, 299; G. Forster, "Geschichte der Englischen Litteratur," AA VII, 235, 237.

161. M. Forkel to Voss, June 28, 1791, AA XVI, 538.

162. Ibid., September 24, 1791, AA XVI, 564.

163. G. Forster, *Voyage round the World* (2000), 199; G. Forster, "Vorrede zur deutschen Übersetzung [of Thomas Paine's *Die Rechte des Menschen*], AA VIII, 222.

164. G. Forster to Dohm, July 13–18, 1790, AA XVI, 158; G. Forster to C. G. Heyne, July 13, 1790, AA XVI, 157; G. Forster, "Erinnerungen aus dem Jahr 1790," AA VIII, 287. Rolf Reichardt points to the influence of scenes like this on Forster's development—scenes that, as a writer, he sought to create in turn for his reader; cf. Reichardt, "Die visualisierte Revolution."

165. G. Forster to Dorsch, January 17, 1793, AA XVII, 308.

166. Dumont provides an interim balance of the early phase of revolutionization in *Die Mainzer Republik*, 252–257.

167. G. Forster, "Über das Verhältnis der Mainzer," AA X/1, 28.

168. Dumont, *Die Mainzer Republik*, 296.

169. For the transition to French occupation politics as of December 15, 1792, cf. Dumont, *Die Mainzer Republik*, 259; esp. 288.

170. "Unterricht für die Gemeineversammlungen, und die in den Städten ein-

zurichtenden Urversammlungen, nebst einem Anhang von den Verrichtungen der Municipalitäten," AA X/1, 155.

171. Soemmerring to C. G. Heyne, January 15, 1793, in Soemmerring, *Werke*, 20:78; Laukhard, *Laukhards Leben und Schicksale*, 338. The third part contains events, experiences, and commentary during the field campaign against France from the start to the blockade of Landau.

172. This information is from Neugebauer-Wölk, "Das Rote und das Schwarze Buch."

173. G. Forster, "Darstellung der Revolution in Mainz," AA X/1, 557.

174. According to Neugebauer-Wölk, "Das Rote und das Schwarze Buch," 64.

175. G. Forster, "Darstellung der Revolution in Mainz," AA X/1, 557.

176. G. Forster to T. Forster, March 14, 1793, AA XVII, 331.

177. Ibid., March 21, 1793, AA XVII, 334.

178. Communication from March 16, 1793, signed by Forster as vice president of general administration, AA X/1, 166.

179. G. Forster to T. Forster, March 21, 1793, AA XVII, 334; decree by the Rhenish-German National Convention, convened in Mainz on March 18, 1793, articles 1 and 2, quoted in Dumont, *Die Mainzer Republik*, 426.

180. G. Forster to T. Forster, March 21, 1793, AA XVII, 334; Soemmerring to C. G. Heyne, March 19, 1793, in Soemmerring, *Werke*, 20:88.

181. Soemmerring to C. G. Heyne, January 29, 1793, in Soemmerring, *Werke*, 20:81.

182. G. Forster, "Rede über die Vereinigung des rheinisch-deutschen Freistaats mit der Frankenrepublik," AA X/1, 463.

183. Ibid., 466.

184. Address to the Paris National Convention, given on March 23–24, 1793, and signed by ninety deputies on March 25, 1793, AA X/1, 468–470. Cf. Dumont, *Die Mainzer Republik*, 436; 504 lists only eighty-nine signers.

185. G. Forster to T. Forster, March 25, 1793, AA XVII, 335.

186. C. G. Heyne to Soemmerring, August 27, 1792, in Soemmerring, *Werke*, vol. 19/2, 879; Cf. Seibt, *Mit einer Art von Wut*.

187. G. Forster, "Über lokale und allgemeine Bildung," AA VII, 55; G. Forster, "Reise von Kassel nach Wilna 1784," *Tagebücher*, AA XII, 128.

188. The simple contrast that Karol Sauerland draws is thus inadequate: Sauerland, "Zwei Revolutionsauffassungen." Steiner offers a biographical sketch that sounds out neither the commonalities nor the tensions of Forster and Goethe's encounters in "'Uns hat zu Männern geschmiedet die allmächtige Zeit.'"

189. For Goethe's three tours of Switzerland, see the wonderful book by Adolf Muschg, *Von einem, der auszog, leben zu lernen: Goethes Reisen in die Schweiz* (Frankfurt am Main: Suhrkamp, 2004).

190. Cf. Goldstein, *Die Entdeckung der Natur*, 176.

191. G. Forster to Jacobi, October 10, 1779, AA XIII, 248.

192. Ibid.

193. Ibid., November 2, 1779, AA XIII, 252.

194. Ibid., October 10, 1779, AA XIII, 248; G. Forster to J. R. Forster, October 24, 1779, AA XIII, 250.

195. G. Forster to Jacobi, October 10, 1779, AA XIII, 248.

196. G. Forster to J. R. Forster, October 24, 1779, AA XIII, 251.

197. G. Forster to Jacobi, November 13, 1783, AA XIII, 503; March 9, 1784, AA XIV, 32.

198. G. Forster to F. L. W. Meyer, September 14, 1785, AA XIV, 362; G. Forster to C. G. Heyne, September 19, 1785, AA XIV, 363; G. Forster to Meyer, September 14, 1785, AA XIV, 362.

199. Goethe, "Campaign in France," 5:619.

200. Goethe, "Gesamtschema von 1810 zu: Campagne in Frankreich 1792," in Goethe, *Sämtliche Werke*, XIV, 590.

201. Goethe to G. Forster, June 25, 1792, AA XVIII, 540; G. Forster to C. G. Heyne, October 27, 1792, AA XVII, 223.

202. G. Forster to Jacobi, April 6, 1792, AA XVII, 92; G. Forster to C. G. Heyne, April 7, 1792, AA XVII, 92; G. Forster to Voss, April 14, 1792, AA XVII, 94; Goethe, "Campaign in France 1792," 5:619.

203. J. W. v. Goethe, *Zur Morphologie II, I (Bedeutende Fördnis durch ein einziges geistreiches Wort)*, in Goethe, *Sämtliche Werke*, XII, 308.

204. Goethe to Schiller, March 9, 1802., in Schmitz, *Correspondence between Schiller and Goethe*, 407.

205. J. W. v. Goethe, *Des Epimenides Erwachen*, in Goethe, *Sämtliche Werke*, XI/2, 155.

206. G. Forster, "Parisische Umrisse," AA X/1, 595–596.

207. Goethe, in a letter of June 22, 1781, to Johann Caspar Lavater, quoted in Nietzsche, *Pre-Platonic Philosophers*, 109.

208. Goethe to Lavater, June 22, 1781, in Goethe, *Werke* (Weimar), part 4, 5:149.

209. G. Forster to T. Forster, April 16, 1793, AA XVII, 344; June 10, 1793, AA XVII, 365.

210. Goethe, conversation with Eckermann, April 27, 1825, in *Conversations of Goethe with Eckermann*, 139; Goethe, *Goethe's Theory of Colours*, xxiv.

211. Goethe, "Siege of Mainz," 5:758.

212. Ibid., 761, 763, 758; Goethe, *Italian Journey*, 193.

213. Goethe, "Siege of Mainz," 5:776, 762.

214. Ibid., 5:757.

215. Ibid., 5:765; G. Forster, *Voyage round the World* (2000), 200; Goethe, "Siege of Mainz," 5:771, 767. Goethe's relationship to violence during the recapturing of Mainz is ambivalent. On the one hand, he shows understanding of the attacks—"the capturing up from below," that is, by the common people, "strikes me as good," he writes to Jacobi, July 27, 1793 (in Goethe, *Werke* [Weimar], part 4, 10:101). On the other hand, he sets a limit to violence through his actual or later fictionalized courageous intervention, as he describes it in the "Siege of Mainz" of July 25, 1793. Cf. the lucid study by Seibt, *Mit einer Art von Wut.*

216. Goethe to C. G. Voigt, July 3, 1793, in Goethe, *Werke* (Weimar), part 4, 10:84.

217. Cf. Goethe's account "Siege of Mainz," 5:762.

218. Goethe, "Campaign in France," 5:651–652.

V

1. J. W. v. Goethe to S. T. Soemmerring, December 5, 1793, in Goethe, *Werke* (Weimar), Abt. IV, v. 10, 130; G. Forster to T. Forster, April 5, 1793, AA XVII, 338.

2. Dippel, "Forster und die angebliche Reichsacht," 253; G. Forster to T. Forster, January 2, 1793, AA XVII, 296.

3. I thank him for referring me to his contribution: Schöne, "Unter aller Würde der Societät," 441.

4. Ibid. I have not carried over the emphasis indicated in the original.

5. Cf. G. Steiner, "Jacobiner und Societät der Wissenschaften."

6. G. Forster to T. Forster, August 21, 1793, AA XVII, 428; July 7, 1793, AA XVII, 383.

7. Ibid., May 4, 1793, AA XVII, 351; June 17, 1793, AA XVII, 370; G. Forster to C. G. Heyne, December 10, 1791, AA XVI, 394.

8. G. Forster to T. Forster, June 17, 1793, AA XVII, 370.

9. Ibid., August 11, 1793, AA XVII, 417.

10. Ibid., May 23, 1793, AA XVII, 360; May 4, 1793, AA XVII, 351.

11. G. Forster, "Parisische Umrisse," AA X/1, 593.

12. G. Forster to T. Forster, April 8, 1793, AA XVII, 341.

13. Ibid., April 13, 1793, AA XVII, 344, 342.

14. G. Forster, "Rede im Pariser Nationalkonvent am 30. März, 1793," AA X/1, 477.

15. G. Forster to T. Forster, April 8, 1793, AA XVII, 341.

16. Ibid., April 13, 1793, AA XVII, 342.

17. Ibid., April 16, 1793, AA XVII, 344; August 21, 1793, AA XVII, 424, 423; October 8, 1793, AA XVII, 459.

18. Ibid., April 16, 1793, AA XVII, 346.

19. As Forster mentions in a letter to T. Forster and L. F. Huber of November 27, 1793, AA XVII, 482.

20. According to Darnton, *Glänzende Geschäfte*, 9.

21. G. Forster to T. Forster, April 27, 1793, AA XVII, 348; June 4, 1793, AA XVII, 362; August 26, 1793, AA XVII, 431.

22. Ibid., June 10, 1793, AA XVII, 366; cf. also ibid., April 27, 1793, AA XVII, 348. Forster repeatedly expressed an interest in India; cf. Esleben, "Forster und Indien."

23. G. Forster to T. Forster, June 10, 1793, AA XVII, 366; June 4, 1793, AA XVII, 363.

24. Ibid., April 8, 1793, AA XVII, 342; June 2, 1793, AA XVII, 362; June 26, 1793, AA XVII, 376.

25. Quoted in Soboul, *Die Große Französische Revolution*, 352.

26. According to the calculations of historian Don Greer, reported in Mazauric, "Terreur," 1023.

27. Furet, "Die Schreckensherrschaft," 1:202.

28. Robespierre, "Über die Grundsätze," 564; Robespierre, "The Political Philosophy of Terror" (a speech to the convention), February 5, 1794.

29. G. Forster to T. Forster, April 27, 1793, AA XVII, 349; Reinalter, "Johann Georg Forsters Revolutionsverständnis," 212; G. Forster to T. Forster, January 28, 1793, AA XVII, 323; Soemmerring to C. G. Heyne, March 19, 1793, in Soemmerring, *Werke*, 20:88.

30. G. Forster to T. Forster, June 10, 1793, AA XVII, 364; March 31, 1793, AA XVII, 337; September 6, 1793, AA XVII, 439.

31. Ibid., August 1, 1793, AA XVII, 410.

32. Ibid., September 12, 1793, AA XVII, 444.

33. G. Forster, "Über die Beziehung der Staatskunst auf das Glück der Menschheit," AA X/1, 589.

34. G. Forster to T. Forster, April 27, 1793, AA XVII, 350; G. Forster to T. Forster and Huber, November 27, 1793, AA XVII, 480; G. Forster to T. Forster, September 30, 1793, AA XVII, 452; September 6, 1793, AA XVII, 440.

35. G. Forster to T. Forster, September 6, 1793, AA XVII, 437; April 8, 1793, AA XVII, 341; September 18, 1793, AA XVII, 446.

36. Börckel, *Adam Lux*, 13.

37. G. Forster to T. Forster, July 19, 1793, AA XVII, 395.

38. Cf. Dumont, "'Sein Leben dem Wahren widmen.'"

39. A. Lux, *Charlotte Corday* (pamphlet of July 19, 1793), in Börckel, *Adam Lux*, 54; G. Forster to T. Forster, July 24, 1793, AA XVII, 404.

40. A. Lux to Suadet and Pétion, June 6, 1793, in Börckel, *Adam Lux*, 21, 26, 27.

41. Lux, *Charlotte Corday*, 55.

42. Ibid., 57.

43. G. Forster to T. Forster, July 23, 1793, AA XVII, 402; July 8, 1793, AA XVII, 385.

44. Ibid., August 21, 1793, AA XVII, 427; Börckel, *Adam Lux*, 1.

45. George Kerner's report, written by his brother J. Kerner, *Das Bilderbuch*, 92; Börckel, *Adam Lux*, 82.

46. G. Forster to T. Forster, November 9–10, 1793, AA XVII, 466; November 20, 1793, AA XVII, 478.

47. G. Forster to T. Forster and Huber, November 27, 1793, AA XVII, 481.

48. G. Forster to Huber, November 15, 1793, AA XVII, 474.

49. G. Forster to T. Forster, July 19, 1793, AA XVII, 395.

50. G. Forster to Huber, November 15, 1793, AA XVII, 474.

51. G. Forster to T. Forster, June 17, 1793, AA XVII, 370.

52. G. Forster, "Über die Beziehung der Staatskunst," AA X/1, 574.

53. Ibid., 565, 566.

54. Ibid., 584.

55. Ibid., 587, 573, 576.

56. Ibid., 568.

57. Ibid., 591.

58. Ibid., 574, 581.

59. Ibid., 591.

60. Ibid., 575.

61. Ibid., 576.

62. Ibid., 576, 591. Echoes of Goethe's *Prometheus* can be heard when Forster says that man is here "to suffer pain, to weep."

63. G. Forster to T. Forster, September 18, 1793, AA XVII, 447; August 1, 1793, AA XVII, 409.

64. Ibid., April 16, 1793, AA XVII, 345.

65. Ibid.

66. Ibid., April 5, 1793, AA XVII, 338.

67. Ibid., June 26, 1793, AA XVII, 376.

68. Burke, *Reflections on the Revolution*, 448, 431, 681, 429.

69. Liesegang, "Das Skandalon der Revolution," 512.

70. Stammen, "Reisen in die Hauptstadt der Revolution," 302.

71. G. Forster to T. Forster, October 8, 1793, AA XVII, 459.

72. Stammen, "Reisen in die Hauptstadt der Revolution," 301; G. Forster to T. Forster, October 8, 1793, AA XVII, 459.

73. G. Forster, "Parisische Umrisse," AA X/1, 593.

74. Ibid., 633, 593.

75. Ibid., 601.

76. Ibid., 609.

77. Ibid., 600.

78. Ibid., 595.

79. Ibid., 596.

80. Ibid., 637.

81. Ibid., 597, 604, 602, 621, 600.

82. Ibid., 600. "Welcome, revolution, with all your evils and abominations!," Forster writes in a letter to his wife of November 9–10, 1793, AA XVII, 466. For an apology for revolutionary violence by Forster, cf. Gilli, "Die Grenzen der Demokratie."

83. G. Forster, "Parisische Umrisse," AA X/1, 600, 614.

84. Ibid., 622.

85. Ibid., 596.

86. Ibid., 619.

87. Ibid., 622.

88. G. Forster to T. Forster, June 26, 1793, AA XVII, 376; G. Forster, "Geschichte der Englischen Litteratur, vom Jahr 1788," AA VII, 80–82; Gibbon, *History of the Roman Empire*, 1:81; G. Forster, "Geschichte der Englischen Litteratur," AA VII, 81.

89. G. Forster to T. Forster, June 26, 1793, AA XVII, 376; G. Forster, "Parisische Umrisse," AA X/1, 623.

90. G. Forster to T. Forster, September 25, 1793, AA XVII, 448; G. Forster to Huber, November 15, 1793, AA XVII, 473; G. Forster, "Über die Beziehung der Staatskunst," AA X/1, 589; G. Forster to T. Forster, January 2, 1793, AA XVII, 295.

91. G. Forster, *Cook, the Discoverer*, 258.

92. G. Forster, "Parisische Umrisse," AA X/1, 595.

93. Ibid.

94. G. Forster to Huber, November 15, 1793, AA XVII, 471; November 11, 1793, AA XVII, 469.

95. G. Forster to F. H. Jacobi, May 18, 1792, AA XVII, 117; G. Forster to T. Forster and Huber, December 27, 1793, AA XVII, 496; Harpprecht, *Georg Forster*, 605; G. Forster to T. Forster and Huber, December 27, 1793, AA XVII, 497.

96. G. Forster to T. Forster, November 6, 1793, AA XVII, 462; G. Forster to T. Forster and Huber, December 20, 1793, AA XVII, 491.

97. G. Forster to T. Forster and Huber, December 11, 1793, AA XVII, 485; December 14, 1793, AA XVII, 487.

98. Ibid., December 19, 1793, AA XVII, 490.

99. Lepenies, "Forster als Anthropologe," 149.

100. G. Forster to T. Forster and Huber, December 2, 1793, AA XVII, 483; December 14, 1793, AA XVII, 487; December 28, 1793, AA XVII, 498; January 4, 1794, AA XVII, 499; G. Forster, "Leitfaden zu einer Geschichte der Menschheit," AA VIII, 193.

101. G. Forster to T. Forster, April 27, 1793, AA XVII, 349; G. Forster to T. Forster and Huber, December 22, 1793, AA XVII, 493; January 2, 1794, AA XVII, 495.

102. G. Forster to T. Forster and Huber, January 4, 1794, AA XVII, 499.

103. S. La Roche to Soemmerring, January 29, 1794, in Soemmerring, *Werke*, 20: 159; *Gazette nationale* (Paris), January 18, 1794, quoted in Enzensberger, *Georg Forster*, 293. For the shifting history of reception after Forster's death until the publication of the Akademie edition of his writings, cf. Scheuer, "'Apostel der Völkerfreiheit.'" For the history of reception in the nineteenth century, cf. Uhlig, "Zwischen Politik, Belletristik und Literaturwissenschaft." See also Peitsch's essential reappraisal of Forster's history of reception from Schlegel to Ulrich Enzensberger, Klaus Harpprecht, and Gerhart Pickerodt, in *Georg Forster*.

104. C. G. Heyne to Soemmerring, January 31, 1794, in Soemmerring, *Werke*, 20: 162; G. C. Lichtenberg to Soemmerring, June 5, 1795, in Soemmerring, *Werke*, 20:221.

105. J. W. v. Goethe, *Herrmann und Dorothea*, in Goethe, *Sämtliche Werke*, IV/1, 592.

106. Ibid., 599.

107. Ibid., 627.

108. Klaus Harpprecht believes it is Forster; Gustav Seibt believes it is Adam Lux: Harpprecht, *Georg Forster*, 27; Seibt, *Mit einer Art von Wut*, 142–145.

109. Goethe, *Herrmann und Dorothea*, in Goethe, *Sämtliche Werke*, IV/1, 627; Goethe to Soemmerring, February 17, 1794, in Goethe, *Werke* (Weimar), Abt. IV, v. 10, 142.

110. Goethe, *Herrmann und Dorothea*, in Goethe, *Sämtliche Werke*, IV/1, 599; Boyle, *Goethe*, 2:544.

111. Goethe, "Siege of Mainz," 5:762.

112. J. W. v. Goethe, *Materialien zur Geschichte der Farbenlehre*, v. 2, in Goethe, *Sämtliche Werke*, X, 913.

113. Heine, *Religion and Philosophy in Germany*, 144.

114. Leitzman, *Georg Forster*, 31; Laukhard, *Laukhards Leben und Schicksale*, 338.

115. Humboldt, *Cosmos*, 70.

116. Uhlig, "Georg Forsters Horizont," 6; Arendt, *On Revolution*, 49; Habermas, *Structural Transformation of Public Sphere*, 93, 101.

117. Habermas, "Natural Law and Revolution."

118. G. Forster, "Erinnerungen aus dem Jahr 1790 in historischen Gemälden und Bildnissen von D. Chodowiecki, D. Berger, C. Kohl, J. F. Bolt, und J. S. Ringck," AA VIII, 285.

119. G. Forster to T. Forster, June 26, 1793, AA XVII, 379.

120. G. Forster, "Vorlesungen über allgemeine Naturkenntnis," AA VI/2, 1754.

English-Language Works

Arendt, Hannah. *On Revolution*. New York: Penguin, 1990.

Benjamin, Walter. "Paris, Capital of the Nineteenth Century." In *The Arcades Project*, translated by Howard Eiland and Kevin McLaughlin, 3–26. Cambridge: Harvard University Press, 1999.

Blumenberg, Hans. *Paradigms for a Metaphorology*. Translated by Robert Savage. Ithaca, NY: Cornell University Press, 2010.

Bodin, Jean. *Six Books of the Commonwealth*. Translated by M. J. Tooley. Oxford: B. Blackwell, 1955.

Bougainville, Louis-Antoine de. *A Voyage Round the World*. Translated by Johann Reinhold Forster. Dublin: J. Exshaw, 1772. Accessed via Gale Cengage Learning, http://find.galegroup.com.

Burke, Edmund. *Reflections on the Revolution in France and Other Writings*. Edited by Jesse Norman. New York: Knopf, 2015.

Chamisso, Adelbert von. *A Voyage around the World with the Romanzov Exploring Expedition in the Years 1815–1818*. Translated and edited by Henry Kratz. Honolulu: University of Hawaii Press, 1986.

Craig, Gordon Alexander. *The Politics of the Unpolitical: German Writers and the Problem of Power, 1770–1871*. Oxford: Oxford University Press, 1995.

Darwin, Charles. *The Voyage of the Beagle*. London: J. M. Dent, Dutton, 1906. Accessed via HathiTrust Digital Library, https://babel.hathitrust.org.

Eliot, George. "German Wit: Henry Heine." In *The Essays of George Eliot*, 99–140. New York: Funk & Wagnalls, 1883.

Ferguson, Adam. *An Essay on the History of Civil Society*. Edited by Fania Oz-Salzberger. Cambridge: Cambridge University Press, 1995.

Forster, Georg. *A Voyage round the World*. Edited by Nicholas Thomas and Oliver Berghof. 2 vols. Honolulu: University of Hawaii Press, 2000.

———. *Cook, the Discoverer*. Sydney: Hordern House, 2007.

———. "New Holland and the British Colony at Botany Bay." Translated by Robert J. King. www.australiaonthemap.org.au/new-holland-and-the-british-colony-at -botany-bay-2/, April 18, 2008.

———. "Something More about the Human Races." In Mikkelsen, *Kant and the Concept of Race*, 143–167.

Forster, Johann Reinhold. *1729–1798: Observations Made During a Voyage Round the World, On Physical Geography, Natural History, And Ethic Philosophy: Especially On: 1. The Earth And Its Strata; 2. Water And the Ocean; 3. The Atmosphere; 4. The Changes of the Globe; 5. Organic Bodies; And 6. The Human Species*. London: G. Robinson, 1778.

Francke, Kuno, William Guild Howard, and Isidore Singer, eds. *The German Classics: Masterpieces of German Literature: Translated into English*. Vol. 2. Translated by Frances H. King and Louis H. Gray. New York: German Publication Society, 1913.

Gibbon, Edward. *The History of the Decline and Fall of the Roman Empire*. Vol. 1. London: Oxford University Press, 1907. Accessed via Google Books, https://books.google .com/books.

Goethe, Johann Wolfgang von. "Campaign in France, 1792." In *From My Life: Poetry and Truth*, edited by Thomas P. Saine and Jeffrey L. Sammons and translated by Thomas P. Saine, 5:614–748. New York: Suhrkamp, 1987.

———. *Conversations of Goethe with Eckermann and Soret*. Translated by John Oxenford. London: G. Bell, 1883. Accessed via HathiTrust Digital Library, https://babel .hathitrust.org.

———. "Goethe's Correspondence with K. F. Zelter." In Francke, Howard, and Singer, *German Classics*, 492–506. Accessed via Google Books, https://books.google.com /books.

———. "Goethe's Correspondence with Wilhelm von Humboldt and His Wife." In Francke, Howard, and Singer, *German Classics*, 469–491. Accessed via Google Books, https://books.google.com/books.

———. *Goethe's Theory of Colours*. Translated by Charles Lock Eastlake. London: John Murray, 1840. Accessed via HathiTrust Digital Library, https://babel.hathitrust .org.

———. *Italian Journey, 1786–1788*. Translated by W. H. Auden and Elizabeth Mayer. London: Penguin, 1970.

———. "Letter to Johann Caspar Lavater, from June 22, 1781." In *The Pre-Platonic Philosophers*, by Friedrich Nietzsche. Translated by Greg Whitlock. Urbana: University of Illinois Press, 2001.

———. *The Poet and the Age*: Vol. 2, *Revolution and Renunciation (1790–1803)*. Translated by Nicholas Boyle. Oxford: Oxford University Press, 2003.

———. "Siege of Mainz." In *From My Life: Poetry and Truth*, edited by Thomas P. Saine and Jeffrey L. Sammons, translated by Thomas P. Saine, 5:749–776. New York: Suhrkamp, 1987.

Habermas, Jürgen. *Between Facts and Norms: Contributions to a Discourse Theory of Law and Democracy*. Translated by William Rehg. Cambridge, MA: MIT Press, 1996.

———. "Natural Law and Revolution." In *Theory and Practice*, translated by John Viertel, 82–120. Boston: Beacon Press, 1973.

———. *The Structural Transformation of the Public Sphere: An Inquiry into a Category of Bourgeois Society*. Translated by Thomas Burger. Cambridge: Polity Press, 1989.

Heine, Heinrich. *Religion and Philosophy in Germany: A Fragment*. Translated by John Snodgrass. Albany: State University of New York Press, 1986.

Herder, Johann Gottfried. "Ideas for a Philosophy of History." In *J. G. Herder on Social and Political Culture*, translated and edited by F. M. Barnard, 255–326. Cambridge: Cambridge University Press, 1969.

———. "Reflections on the History of Greece." In *Outlines of a Philosophy of the History of Man*, translated by T. Churchill, 185–194. London: J. Johnson, 1803. Accessed via Google Books, https://books.google.com/books.

Hobbes, Thomas. *On the Citizen*. Translated by Richard Tuck and Michael Silverthorne. Cambridge: Cambridge University Press, 1998.

Horace. *Satires, Epistles, and Ars Poetica: With an English Translation*. Translated by Henry Rushton Fairclough. Cambridge, MA: Harvard University Press, 1936.

Humboldt, Alexander von. *Cosmos: Sketch of a Physical Description of the Universe*. Translated by Edward Sabine. London: Longman, Green, and Longmans, 1849.

Kant, Immanuel. *The Conflict of the Faculties*. Translated by Mary J. Gregor. Lincoln: University of Nebraska Press, 1992.

———. *Critique of Judgment*. Translated by J. H. Bernard. North Chelmsford, MA: Courier, 2012.

———. *Critique of Pure Reason*. Translated by J. M. D. Meiklejohn. New York: Colonial Press, 1899.

———. "Determination of the Concept of a Human Race." In Mikkelsen, *Kant and the Concept of Race*, 125–141.

———. "Of the Different Human Races." In Mikkelsen, *Kant and the Concept of Race*, 55–71.

———. "On the Use of Teleological Principles in Philosophy." In Mikkelsen, *Kant and the Concept of Race*, 169–194.

———. *Kant: Political Writings*. Translated by Hans Siegbert Reiss. Cambridge: Cambridge University Press, 1991.

———. *Natural Science*. Translated by Eric Watkins. Cambridge: Cambridge University Press, 2012.

———. *On History*. Edited by Lewis White Beck. Translated by Robert E. Anchor. Indianapolis: Bobbs-Merrill, 1963.

———. *Religion within the Boundaries of Mere Reason*. Translated by Allen Wood. Cambridge: Cambridge University Press, 1998.

Kleist, Heinrich von. "The Chilean Earthquake." In *Selected Writings*, translated by David Constantine, 312–323. Indianapolis: Hackett, 1997.

Koselleck, Reinhart. *Futures Past: On the Semantics of Historical Time*. Translated by Keith Tribe. New York: Columbia University Press, 2005.

Marx, Karl. *The Eighteenth Brumaire of Louis Bonaparte*. New York: Cosimo Books, 2008.

Mikkelsen, Jon M. *Kant and the Concept of Race: Late Eighteenth-Century Writings.* Albany: State University of New York Press, 2013.

Montaigne, Michel de. "On the Cannibals." In *The Essays: A Selection*, translated by M. A. Screech, 79–92. London: Penguin, 1993.

Nietzsche, Friedrich. *The Pre-Platonic Philosophers.* Translated by Greg Whitlock. Urbana: University of Illinois Press, 2001.

Polo, Marco. *The Travels of Marco Polo.* Translated by W. Marsden. New York: Dover, 1993.

Pope, Alexander. *Essay on Man.* Edited by Mark Pattison. Oxford: Clarendon Press, 1871. Accessed via HathiTrust Digital Library, https://babel.hathitrust.org.

Rousseau, Jean-Jacques. *A Discourse on Inequality: On the Origin and Basis of Inequality among Men.* Waiheke, NZ: Floating Press, 2009.

———. *Letter to D'Alembert and Writings for the Theater.* Translated and edited by Allan David Bloom and Christopher Kelly. Hanover, NH: University Press of New England, 2004.

———. "The Social Contract." In *The Social Contract and the First and Second Discourses*, translated and edited by Susan Dunn, 149–254. New Haven, CT: Yale University Press, 2002. Accessed via EBSCOhost, http://web.a.ebscohost.com.

Schiller, Friedrich. "Cabal and Love." In *Cabal and Love: Schiller's Poems and Plays*, translated by Matthew Gregory Lewis, 309–375. London: George Routledge, 1889. Accessed via Google Books, https://books.google.com/books.

———. "The Robbers." In *The Works of Friedrich Schiller: Early Dramas and Romances*, translated by H. G. Bohn, 1–129. London: George Bell, 1901. Accessed via Google Books, https://books.google.com/books.

Schmitz, Dora L., trans. *Correspondence between Schiller and Goethe, from 1794–1805.* Vol. 2. London: G. Bell, 1877–1890. Accessed via HathiTrust Digital Library, https://babel.hathitrust.org.

Sieyès, Emmanuel Joseph. *What Is the Third Estate?* Translated by M. Blondel. London: Pall Mall, 1963.

Wallace, Alfred Russel. *The Malay Archipelago, the Land of the Orang-utan and the Bird of Paradise: A Narrative of Travel, with Studies of Man and Nature.* London: Macmillan, 1886. Accessed via HathiTrust Digital Library, https://babel.hathitrust.org.

Bibliography of the German Edition

This bibliography reflects that published in the German edition of this book: Jürgen Goldstein, *Georg Forster: Zwischen Freiheit und Naturgewalt* (Berlin: Matthes & Seitz, 2015).

I. Georg Forster

Georg Forsters Werke. Sämtliche Schriften, Tagebücher, Briefe, hg. von der Akademie der Wissenschaften der DDR, Berlin 1958 ff., seit 1992 fortgeführt von der Berlin-Brandenburgischen Akademie der Wissenschaften; zitiert als »Akademie-Ausgabe« (AA) mit Band- und Seitenangabe.

AA I   *A Voyage round the World*, bearbeitet von R. L. Kahn, Berlin 1968, zweite, unveränderte Auflage Berlin 1986.

AA II   *Reise um die Welt*, 1. Teil, bearbeitet von G. Steiner, Berlin 1965, zweite, unveränderte Auflage Berlin 1989.

AA III   *Reise um die Welt*, 2. Teil, bearbeitet von G. Steiner, Berlin 1966, zweite, unveränderte Auflage Berlin 1989.

AA IV   *Streitschriften und Fragmente zur Weltreise*. Erläuterungen und Register zu Bd. I-IV, bearbeitet von R. L. Kahn, G. Steiner, H. Fiedler, K.-G. Popp, S. Scheibe, Berlin 1970, zweite, unveränderte Auflage Berlin 1989.

AA V   *Kleine Schriften zur Völker- und Länderkunde*, bearbeitet von H. Fiedler, K.-G. Popp, A. Schneider, Ch. Suckow, Berlin 1985.

AA VI/1   *Schriften zur Naturkunde, 1. Teil*, bearbeitet von K.-G. Popp, Berlin 2003.

AA VI/2   *Schriften zur Naturkunde, 2. Teil*, bearbeitet von K.-G. Popp, Berlin 2003.

AA VII   *Kleine Schriften zu Kunst und Literatur. Sakontala*, berarbeitet von
          G. Steiner, Berlin 1963, zweite, unveränderte Auflage Berlin 1990.

AA VIII   *Kleine Schriften zu Philosophie und Zeitgeschichte*, bearbeitet von
          S. Scheibe, Berlin 1974, zweite Auflage Berlin 1991.

AA IX   *Ansichten vom Niederrhein, von Brabant, Flandern, Holland, England und
          Frankreich im April, Mai und Junius 1790*, bearbeitet von G. Steiner, Berlin
          1958.

AA X/1   *Revolutionsschriften 1792/93.* Reden, administrative Schriftstücke,
          Zeitungsartikel, politische und diplomatische Korrespondenz, Aufsätze,
          bearbeitet von K.–G. Popp, Berlin 1990.

AA XI   *Rezensionen*, bearbeitet von H. Fiedler, Berlin 1977, zweite, berichtigte
          Auflage 1992.

AA XII   *Tagebücher*, bearbeitet von B. Leuschner, Berlin 1973, zweite, berichtigte
          Auflage Berlin 1993.

AA XIII   *Briefe bis 1783*, bearbeitet von S. Scheibe, Berlin 1978.

AA XIV   *Briefe 1784–Juni 1787*, bearbeitet von B. Leuschner, Berlin 1978.

AA XV   *Briefe Juli 1787–1789*, bearbeitet von H. Fiedler, Berlin 1981.

AA XVI   *Briefe 1790–1791*, bearbeitet von B. Leuschner, S. Scheibe, Berlin 1980.

AA XVII   *Briefe 1792–1794 und Nachträge*, bearbeitet von K.-G. Popp, Berlin 1989.

AA XVIII *Briefe an Forster*, bearbeitet von B. Leuschner, S. Scheibe, H. Fiedler,
          K.-G.Popp, A. Schneider, Berlin 1982.

Forster, G., *Reise um die Welt. Illustriert von eigener Hand*, mit einem biographischen
    Essay von K. Harpprecht und einem Nachwort von F. Vorpahl, Frankfurt am Main
    2007.
Forster, G., *James Cook, der Entdecker und Fragmente über Capitain Cooks letzte Reise und
    sein Ende*, hg. und mit einem Nachwort versehen von F. Vorpahl und mit acht Farb-
    tafeln von Forsters eigener Hand, Berlin 2008.

II. Georg-Forster-Bibliographien

Fiedler, H., *Georg-Forster-Bibliographie. 1767–1970*, Berlin 1971.
Klenke, C.-V., »Georg-Forster-Bibliographie 1970–1993«, in *Georg Forster in inter-
    disziplinärer Perspektive*, hg. von C.-V. Klenke, Beiträge des Internationalen Georg
    Forster-Symposions in Kassel, 1. bis 4. April 1993, Berlin 1994, 341–415.

III. Georg-Forster-Studien

Georg-Forster-Studien, hg. im Auftrag der Georg-Forster-Gesellschaft, Bd. I-II, Berlin
    1997 und 1998, ab Bd. III, Kassel 1999 ff.

IV. Historische Quellen: Verwendete Werke und Werkausgaben

Aristoteles, *Politik*, in *Werke in deutscher Übersetzung*, begründet von E. Grumach,
    hg. von H. Flashar, Bde. 9/I-IV, Berlin I-II: 1991, III: 1996, IV: 2005.

Ascher, S., *Ideen zur natürlichen Geschichte der politischen Revolution*, ohne Ort 1802 (Nachdruck Kronberg/Ts. 1975).

Bodin, J., *Sechs Bücher über den Staat*, übersetzt und mit Anmerkungen versehen von B. Wimmer, eingeleitet und hg. von P. C. Mayer-Tasch, 2 Bde., München 1981/1986.

Börckel, A., *Adam Lux, ein Opfer der Schreckenszeit. Nach seinen Schriften und Berichten seiner Zeitgenossen*, Mainz 1892.

Bougainville, L.-A. de, *Reise um die Welt*, hg. von K.-G. Popp, Stuttgart 1980.

Burke, E., *Betrachtungen über die Französische Revolution*, aus dem Englischen übertragen von F. Gentz, hg. von U. Frank-Planitz, Zürich 1987.

Chamisso, A. v., *Reise um die Welt*, mit 150 Lithographien von L. Choris und einem essayistischen Nachwort von M. Glaubrecht, Berlin 2012.

Cook, J., *The Journals of Captain James Cook on his Voyages of Discovery*, edited from the Original Manuscripts by J. C. Beaglehole: Vol. I: *The Voyage of the Endeavour 1768–1771*, Cambridge 1955; Vol. II: *The Voyage of the Resolution and Adventure 1772–1775*, Cambridge 1961, reprinted with Addenda and Corrigenda Cambridge 1969; Vol. III: *The Voyage of the Resolution and Discovery 1776–1780*, Pt. 1–2, Cambridge 1967.

Darwin, Ch., *Die Fahrt der Beagle. Tagebuch mit Erforschungen der Naturgeschichte und Geologie der Länder, die auf der Fahrt von HMS Beagle unter dem Kommando von Kapitän Fitz Roy, RN, besucht wurden*, mit einer Einleitung von D. Kehlmann, dt. von E. Schönfeld, Hamburg 2006.

Eckermann, J. P., *Gespräche mit Goethe in den letzten Jahren seines Lebens*, in J. W. Goethe, *Sämtliche Werke nach Epochen seines Schaffens* (Münchner Ausgabe), München 2006, Bd. XIX.

Elliott, J., »The Memoirs of John Elliott«, in J. Elliott, R. Pickersgill, *Captain Cook's Second Voyage: The Journals of Lieutenants Elliott and Pickersgill*, edited and introduced by Ch. Holmes, London 1984, 1–47.

Ferguson, A., *Versuch über die Geschichte der bürgerlichen Gesellschaft*, hg. und eingeleitet von Z. Batscha und H. Medick, Frankfurt am Main 1988.

Forster, J. R., »Ueber Georg Forster«, in *Annalen der Philosophie und des philosophischen Geistes von einer Gesellschaft gelehrter Männer*, hg. von L. H. Jakob, Erster Jahrgang, Halle/Leipzig 1795, *Philosophischer Anzeiger*: Stück 2 vom 14. Januar, Spalte 9–16, Fortsetzung: Stück 16 vom 15. April, Spalte 121–126.

——, *Beobachtungen während der Cookschen Weltumsegelung 1772–1775. Gedanken eines deutschen Teilnehmers*, unveränderter Neudruck der 1783 erschienenen *Bemerkungen über Gegenstände der physischen Erdbeschreibung, Naturgeschichte und sittlichen Philosophie auf seine Reise um die Welt gesammlet*, mit einer Einführung von H. Beck, Stuttgart 1981.

——, *The Resolution Journal of Johann Reinhold Forster 1772–1775*, edited by M. E. Hoare, 4 vols., London 1982.

——, *Forsters Bilder von der Weltumsegelung mit Cook in der Forschungs- und Landesbibliothek Gotha. Führer zur Ausstellung anläßlich des 200. Todestages von Georg Forster*, Forschungs- und Landesbibliothek Gotha 1994.

Gibbon, E., *Verfall und Untergang des Römischen Reiches*, hg. von D. A. Saunders, aus dem Englischen von J. Sporschil, Frankfurt am Main 2000.

Gilbert, G., *Captain Cook's Final Voyage. The Journal of Midshipman George Gilbert*, introduced and edited by Ch. Holmes, Partridge Green/Horsham (Sussex) 1982.

Goethe, J. W., *Werke*, hg. im Auftrage der Großherzogin Sophie von Sachsen, Weimar 1887–1919, Nachdruck München 1987.

———, *Werke*, hg. von E. Trunz, 2. Auflage, Hamburg 1981.

———, *Sämtliche Werke nach Epochen seines Schaffens*, hg. von K. Richter, in Zusammenarbeit mit H. G. Göpfert, N. Miller und G. Sauder, München 2006.

Grimm, J. und W., *Deutsches Wörterbuch*, Leipzig 1854–1960.

Heine, H., *Historisch-kritische Gesamtausgabe der Werke*, hg. von M. Windfuhr, Hamburg 1973–1997 (Düsseldorfer Ausgabe).

Herder, J. G., *Ideen zur Philosophie der Geschichte der Menschheit*, in *Werke*, Bd. 6, Frankfurt am Main 1989.

Hobbes, Th., *Vom Bürger*, in ders., *Vom Menschen. Vom Bürger. Elemente der Philosophie II/III*, eingeleitet und hg. von G. Gawlick, Hamburg 1994.

Home, H., Lord Kames, *Sketches of the History of Man*, 2nd. ed. 1778, 4 vols. (Reprint: Hildesheim 1968).

Horaz, *Ars Poetica/Die Dichtkunst*, lat./dt., übersetzt und mit einem Nachwort hg. von E. Schäfer, Stuttgart 1972.

Huber, Th., *Briefe*, Bd. I., 1774–1803, Tübingen 1999.

Humboldt, A. v., *Kosmos. Entwurf einer physischen Weltbeschreibung*, ediert und mit einem Nachwort versehen von O. Ette und O. Lubrich, Frankfurt am Main 2004.

Jaucourt, L. de, Art. »Révolution (Moderne Geschichte Englands)«, in *Diderots Enzyklopädie*, mit Kupferstichen aus den Tafelbänden, ediert von A. Selg und R. Wieland, aus dem Französischen von H. Fock, Th. Lücke, E. Moldenhauer und S. Müller, Berlin 2013, 360.

Joubert, J., *Carnets*, Textes recueillis sur les manuscrits autographes par A. Beaunier, Paris 1994.

Kant, I., *Gesammelte Schriften*, hg. von der Königlich Preußischen Akademie der Wissenschaften (fortgeführt von der Berlin-Brandenburgischen Akademie der Wissenschaften), Berlin 1900 ff.

———, *Werkausgabe*, hg. von. W. Weischedel, Frankfurt am Main 1977.

Kerner, J., *Das Bilderbuch aus meiner Knabenzeit. Erinnerungen aus den Jahren 1786 bis 1804*, Braunschweig 1849.

Kleist, H. v., »Das Erdbeben in Chili«, in *Sämtliche Werke und Briefe. Münchner Ausgabe*, auf der Grundlage der Brandenburger Ausgabe, hg. von R. Reuß und P. Staengle, München/Frankfurt am Main 2010, Bd. II, 148–163.

Laukhard, F. C., *F. C. Laukhards Leben und Schicksale, von ihm selbst beschrieben. Dritter Theil, welcher dessen Begebenheiten, Erfahrungen und Bemerkungen während des Feldzugs gegen Frankreich von Anfang bis zur Blokade von Landau enthält*, Leipzig 1796.

Leitzmann, A., *Georg Forster. Ein Bild aus dem Geistesleben des achtzehnten Jahrhunderts*, Halle an der Saale 1893.

Lichtenberg, G. Ch., *Schriften und Briefe*, hg. von W. Promies, Bd. I-IV, München 1967–1972.

Locke, L., *An Essay concerning Human Understanding*, edited by P. H. Nidditch, Oxford 1975.

Marx, K., *Der 18. Brumaire des Louis Bonaparte*, Frankfurt am Main 1965.

Montaigne, M. de, *Essais*, erste moderne Gesamtübersetzung von Hans Stilett, Frankfurt am Main 1998, korrigierte Auflage 1999.

Montesquieu, *De l'esprit des lois*, in *Œuvres complètes*, texte présenté et annoté par R. Caillois (Bibliothèque de la Pléiade), Bd. II, Paris 2001.

Paine, Th., *The Rights of Man*, London 1950.

Polo, M., *Il Milone/Die Wunder der Welt*, Übersetzung aus altfranzösischen und lateinischen Quellen und Nachwort von E. Guignard, Zürich 1983.

Pope, A., *Essay on Man*, engl./dt., übersetzt von E. Breidert, mit einer Einleitung hg. von W. Breidert, Hamburg 1993.

Robespierre, M., »Über die Grundsätze der revolutionären Regierung« (Rede vor dem Konvent vom 25. Dezember 1793), in ders., *Ausgewählte Texte*, deutsch von M. Unruh, mit einer Einleitung von C. Schmid, Gifkendorf ²1989, 562–581.

Rousseau, J.-J., »Abhandlung über den Ursprung und die Grundlagen der Ungleichheit unter den Menschen«, in ders., *Schriften*, hg. von H. Ritter, München/Wien 1978, Bd. 1, 165–302.

———, »Brief an Herrn d'Alembert über seinen Artikel ›Genf‹ im VII. Band der Enzyklopädie und insbesondere über den Plan, ein Schauspielhaus in dieser Stadt zu errichten«, in ders., *Schriften*, hg. von H. Ritter, München/Wien 1978, Bd. I, 333–474.

———, *Vom Gesellschaftsvertrag oder Grundsätze des Staatsrechts*, Stuttgart 2011.

Schiller, F., *Werke. Nationalausgabe*, begründet von J. Petersen, fortgeführt von L. Blumenthal und B. v. Wiese, Weimar 1943 ff.

———, *Sämtliche Werke*, auf der Grundlage der Textedition von H. G. Göpfert hg. von P.-A. Alt, 5 Bde., München/Wien 2004.

Schiller, F./Goethe, J. W., *Der Briefwechsel zwischen Schiller und Goethe*, hg. von Emil Staiger, revidierte Neuausgabe von H.-G. Dewitz, Frankfurt am Main 2005.

Schlegel, F., »Georg Forster. Fragment einer Charakteristik der deutschen Klassiker«, in *Kritische Friedrich-Schlegel-Ausgabe*, hg. von E. Behler, Abteilung I, Bd. II, München/Paderborn/Wien 1967, 78–99.

Sieyès, E. J., *Was ist der Dritte Stand?*, hg. von O. Dann, Essen 1988.

Soemmerring, S. Th., *Werke*, begründet von G. Mann, hg. von J. Benedum und W. F. Kümmel im Auftrag der Akademie der Wissenschaften und der Literatur, Mainz, Stuttgart/Jena et al. 1990 ff.

Wallace, A. R., *Der Malayische Archipel. Die Heimath des Orang-Utan und des Paradiesvogels. Reiseerlebnisse und Studien über Land und Leute*, Berlin 2009.

Wieland, Ch. M., »Auszüge aus Forsters Reise um die Welt«, in *Der Teutsche Merkur*, 1778, 4. Vierteljahr, 137–155.

Zedler, J. H. (Hg.), *Grosses vollständiges Universal Lexicon aller Wissenschafften und*

Künste, welche bißhero durch menschlichen Verstand und Witz erfunden und verbessert worden, Halle/Leipzig 1732–1754 (2. vollständiger photomechanischer Nachdruck: Graz 1993–1999).

V. Verwendete neuere Literatur

Albus, V., *Weltbild und Metapher. Untersuchungen zur Philosophie im 18. Jahrhundert*, Würzburg 2001.

Arendt, H., *Über die Revolution*, München/Zürich 2011.

Baxmann, I., *Die Feste der Französischen Revolution. Inszenierung von Gesellschaft als Natur*, Weinheim/Basel 1989.

Beck, H., »Georg Forster und Alexander von Humboldt. Zur Polarität ihres geographischen Denkens«, in *Der Weltumsegler und seine Freunde. Georg Forster als gesellschaftlicher Schriftsteller der Goethezeit*, hg. von D. Rasmussen, Tübingen 1988, 175–188.

Benjamin, W., »Paris, die Hauptstadt des XIX. Jahrhunderts«, in *Gesammelte Schriften*, Bd. V/1, Frankfurt am Main 1982, 45–59.

Berg, E., *Zwischen den Welten. Über die Anthropologie der Aufklärung und ihr Verhältnis zu Entdeckungs-Reise und Welt-Erfahrung mit besonderem Blick auf das Werk Georg Forsters*, Berlin 1982.

Blumenberg, H., »Paradigmen zu einer Metaphorologie«, in *Archiv für Begriffsgeschichte* 6 (1960), 7–142.

———, *Schiffbruch mit Zuschauer. Paradigma einer Daseinsmetapher*, Frankfurt am Main 1979.

Bödeker, H. E., »Die ›Natur des Menschen so viel als möglich in mehreres Licht [. . .] setzen‹. Ethnologische Praxis bei Johann Reinhold und Georg Forster«, in *Natur—Mensch—Kultur. Georg Forster im Wissenschaftsfeld seiner Zeit*, hg. von J. Garber/T. van Hoorn, Hannover-Laatzen 2006, 143–170.

Börner, K. H., *Auf der Suche nach dem irdischen Paradies. Zur Ikonographie der geographischen Utopie*, Frankfurt am Main 1984.

Boyle, N., *Goethe. Der Dichter in seiner Zeit, Bd. II: 1791–1803*, aus dem Englischen übersetzt von H. Fliessbach, München 1999.

Braudel, F., *Das Mittelmeer und die mediterrane Welt in der Epoche Philipps II.*, 3 Bde., Frankfurt am Main 1990.

Braudel, F./Duby, G./Aymard, M., *Die Welt des Mittelmeeres. Zur Geschichte und Geographie kultureller Lebensformen*, hg. von F. Braudel, aus dem Französischen von M. Jakob, Frankfurt am Main 1990.

Braun, M., *»Nichts Menschliches soll mir fremd sein« — Georg Forster und die frühe deutsche Völkerkunde vor dem Hintergrund der klassischen Kulturwissenschaften*, Bonn 1991.

Bucher, G., *Die Spur des Abendsterns. Die abenteuerliche Erforschung des Venustransits*, Darmstadt 2011.

Dalos, G., *Geschichte der Russlanddeutschen. Von Katharina der Großen bis zur Gegenwart*, deutsche Bearbeitung von E. Zylla, München 2014.

Darnton, R., *Glänzende Geschäfte. Die Verbreitung von Diderots Encyclopédie oder: Wie verkauft man Wissen mit Gewinn?*, aus dem Englischen und Französischen von H. Günther, Berlin 1993.

Demandt, A., *Metaphern für Geschichte. Sprachbilder und Gleichnisse im historisch-politischen Denken*, München 1978.

Dippel, H., »Georg Forster und England«: Weltläufigkeit und Tradition im Denken des Forschers und Revolutionärs, in *Georg-Forster-Studien* I, Berlin 1997, 101–124.

——, »Georg Forster in der deutschen Erinnerungskultur«, in *Georg-Forster-Studien* XI/1, Kassel 2006, 1–29.

——, »Georg Forster und die angebliche Reichsacht«, in *Georg-Forster-Studien* XIII, Kassel 2008, 235–255.

Dumont, F., »Das ›Seelenbündnis‹. Die Freundschaft zwischen Georg Forster und Samuel Thomas Soemmerring«, in *Der Weltumsegler und seine Freunde. Georg Forster als gesellschaftlicher Schriftsteller der Goethezeit*, hg. von D. Rasmussen, Tübingen 1988, 70–100.

——, *Die Mainzer Republik von 1792/1793*, 2. Auflage, Alzey 1993.

——, »Georg Forster als Demokrat. Theorie und Praxis eines deutschen Revolutionärs«, in *Georg-Forster-Studien* I, Berlin 1997, 125–153.

——, »Seeing is believing. Religion und Konfession bei Georg Forster und Samuel Thomas Soemmerring«, in *Georg-Forster-Studien* III, Kassel 1999, 167–196.

——, »›Sein Leben dem Wahren widmen‹. Adam Lux als historische Gestalt«, in *St. Zweig, Adam Lux*, Oldenburg am Main 2003, 113–146.

Enke, U., »›Praelectiones anatomicae‹. Georg Forsters anatomische Studien bei Soemmerring in Kassel«, in *Georg-Forster-Studien* III, Kassel 1999, 1–18.

Enzensberger, U., *Georg Forster. Ein Leben in Scherben*, Frankfurt am Main 1996.

Esleben, J., »Übersetzung als interkulturelle Kommunikation bei Georg Forster«, in *Georg-Forster-Studien* IX, Kassel 2004, 165–179.

——, »Forster und Indien«, in *Georg-Forster-Studien* XI/2, Kassel 2006, 407–426.

Ewert, M., »Vernunft, Gefühl und Phantasie im schönsten Tanze vereint«, in *Die Essayistik Georg Forsters*, Würzburg 1993.

Fagot, P., »In den eigenen Grenzen exiliert: Georg Forster in Polen (1784–1787)«, in *Georg-Forster-Studien* IX, Kassel 2004, 247–261.

Fagot, P., »Polen als Georg Forsters Gegenstück zu Tahiti«, in *Georg-Forster-Studien* XI/2, Kassel 2006, 595–610.

Fink, G.-L., »Klima und Kulturtheorien der Aufklärung«, in *Georg-Forster-Studien* II, Berlin 1998.

Fischer, R., *Reisen als Erfahrungskunst. Georg Forsters »Ansichten vom Niederrhein«. Die »Wahrheit« in den »Bildern des Wirklichen«*, Frankfurt am Main 1990.

——, »›Wer ist der hohe Fremdling in dieser Hülle (. . .)?‹. Georg Forsters Kunstbetrachtungen zwischen Klassizismus, Klassik und Romantik«, in *Georg-Forster-Studien* VI, Kassel 2006, 25–49.

Franke, E. A., »Über Georg Forsters Krankheiten und Tod«, in *Georg-Forster-Studien* IV, Kassel 2000, 187–220.

Fritscher, B., »Vulkanismusstreit und Französische Revolution. Gedanken zu einer zeitlichen Parallele«, in *Geosciences/Geowissenschaften. Proceedings of the Symposium of the XVIIIth International Congress of History of Science at Hamburg-Munich, 1.-9. August 1989*, hg. von M. Büttner/E. Kohler, III. Teil, Bochum 1991, 169–185.

———, »Die Entmoralisierung der Naturgewalten: Vulkane und politische Revolutionen im System der Natur«, in *Elementare Gewalt. Kulturelle Bewältigung. Aspekte der Naturkatastrophe im 18. Jahrhundert* (Jahrbuch der Österreichischen Gesellschaft zur Erforschung des achtzehnten Jahrhunderts, 14.-15. Bd.), hg. von F. Eybl/H. Heppner/A. Kernbauer, Wien 2000, 217–237.

Furet, F., »Art. ›Die Schreckensherrschaft‹«, in *Kritisches Wörterbuch der Französischen Revolution*, hg. von F. Furet und M. Ozouf, 2 Bde., Frankfurt am Main 1996, Bd. I, 193–215.

Garber, J., »Anthropologie und Geschichte. Spätaufklärerische Staats- und Geschichtsdeutung im Metaphernfeld von Mechanismus und Organismus«, in *Georg Forster in interdisziplinärer Perspektive*, hg. von C.-V. Klenke, Berlin 1994, 193–210.

———, »Reise nach Arkadien. Bougainville und Georg Forster auf Tahiti«, in *Georg-Forster-Studien* I, Berlin 1997, 19–50.

Geulen, Ch., *Geschichte des Rassismus*, München 2007.

Gilli, M., »Reform und Revolution bei Georg Forster«, in *Georg-Forster-Studien* IV, Kassel 2000, 17–61.

———, »Die Grenzen der Demokratie: Die Gewalt in den ›Parisischen Umrissen‹«, in *Georg-Forster-Studien* VIII, Kassel 2003, 219–235.

———, »Das Politische in Forsters Rezensionen«, in *Georg-Forster-Studien* IX, Kassel 2004, 225–246.

———, »Auf dem Weg von der Wissenschaft zur Philosophie und Politik: ›Ueber Leckereyen‹ und ›Ueber die Schädlichkeit der Schnürbrüste‹«, in *Georg-Forster-Studien* XIII, Kassel 2008, 61–72.

———, »Georg Forster als Kunstkritiker«, in *Georg-Forster-Studien* XVI, Kassel 2011, 69–86.

Glaubrecht, M., *Am Ende des Archipels. Alfred Russel Wallace*, Berlin 2013.

Godel, R./Stiening, G. (Hg.), *Klopffechtereien—Missverständnisse—Widersprüche? Methodische und Methodologische Perspektiven auf die Kant-Forster-Kontroverse*, Paderborn 2012.

Goldstein, J., *Die Entdeckung der Natur. Etappen einer Erfahrungsgeschichte*, Berlin 2013.

Gordon, J. St., »Reinhold and Georg Forster in England, 1766–1780«, Dissertation, Duke University 1975.

Graczyk, A., »Forschungsreise und Naturbild bei Georg Forster und Alexander von Humboldt«, in *Georg-Forster-Studien* VI, Kassel 2001, 89–116.

Gribbin, J., *The Fellowship. The Story of a Revolution*, London 2005.

Habermas, J., »Naturrecht und Revolution«, in ders., *Theorie und Praxis. Sozialphilosophische Studien*, Frankfurt am Main 1978, 89–127.

———, *Strukturwandel der Öffentlichkeit. Untersuchungen zu einer Kategorie der bürgerlichen Gesellschaft*, Frankfurt am Main 1990.

————, *Faktizität und Geltung. Beiträge zur Diskurstheorie des Rechts und des demokratischen Rechtsstaats*, Frankfurt am Main 1992.

Harpprecht, K., *Georg Forster oder Die Liebe zur Welt. Eine Biographie*, Reinbek bei Hamburg 1987.

Hauser-Schäublin, B., »Das Leben an Bord«, in *James Cook und die Entdeckung der Südsee. Katalog zur Ausstellung der Kunst- und Ausstellungshalle der Bundesrepublik Deutschland*, Bonn, München 2009, 83–86.

Henning, H., »›Vortrefflicher Mann‹ und ›bester Freund‹. Herders Begegnung mit Georg Forster«, in *Der Weltumsegler und seine Freunde. Georg Forster als gesellschaftlicher Schriftsteller der Goethezeit*, hg. von D. Rasmussen, Tübingen 1988, 21–58.

Hoare, M.E., *The Tactless Philosopher. Johann Reinhold Forster (1729–98)*, Melbourne 1975.

Hochadel, O., »Natur—Vorsehung—Schicksal. Zur Geschichtsteleologie Georg Forsters«, in *Wahrnehmung—Konstruktion—Text. Bilder des Wirklichen im Werk Georg Forsters*, hg. von J. Garber, Tübingen 2000, 77–104.

Hocks, P./Schmidt, P., *Literarische und politische Zeitschriften 1789–1805. Von der politischen Revolution zur Literaturrevolution*, Stuttgart 1975.

Holdenried, M., »Erfahrene Aufklärung: Philosophische Reisen in zerstörte Idyllen. Georg Forster als philosophischer Reisender: *Reise um die Welt* (1777)«, in *Georg-Forster-Studien* XI/1, Kassel 2006, 131–145.

Hoorn, T. v., »Zwischen Humanitätsideal und Kunstkritik: Georg Forsters Kunstansichten«, in *Georg-Forster-Studien* V, Kassel 2000, 103–126.

————, »Physische Anthropologie und normative Ästhetik. Georg Forsters kritische Rezeption der Klimatheorie in seiner *Reise um die Welt*«, in *Georg-Forster-Studien* VIII, Kassel 2003, 139–161.

————, »Dem Leibe abgelesen«. *Georg Forster im Kontext der physischen Anthropologie des 18. Jahrhunderts*, Tübingen 2004.

Ikni, G.-R., Art. »Grande Peur«, in *Dictionnaire historique de la Révolution française*, hg. von A. Soboul, Paris 1989, 517 f.

Jahn, I., »›Scientia Naturae—Naturbetrachtung oder Naturwissenschaft?‹ Georg Forsters Erkenntnisfragen zu biologischen Phänomenen in Vorlesungs-Manuskripten aus Wilna und Mainz (1786–1793)«, in *Georg Forster in interdisziplinärer Perspektive*, hg. von C.-V. Klenke, Berlin 1994, 159–177.

Joppien, R., »Georg Forster und William Hodges—Zeugnisse einer gemeinsamen Reise um die Welt«, in *Georg Forster in interdisziplinärer Perspektive*, hg. von C.-V. Klenke, Berlin 1994, 77–102.

Kittsteiner, H. D., »Kants Theorie des Geschichtszeichens. Vorläufer und Nachfahren«, in ders. (Hg.), *Geschichtszeichen*, Köln/Weimar/Wien 1999, 81–115.

Koch, P., »Selbstbildung und Leserbildung. Zu Form und gesellschaftlicher Funktion der ›Ansichten vom Niederrhein‹«, in *Georg Forster in seiner Epoche*, hg. von G. Pickerodt, Berlin 1982, 8–39.

Kohl, K.-H., *Entzauberter Blick. Das Bild vom Guten Wilden und die Erfahrung der Zivilisation*, Frankfurt am Main 1986.

Koselleck, R., »Revolution als Begriff und als Metapher. Zur Semantik eines einst emphatischen Worts«, in ders., *Begriffsgeschichten. Studien zur Semantik und Prag-matik der politischen und sozialen Sprache*, Frankfurt am Main 2006, 240–251.

Kronauer, U., »Rousseaus Kulturkritik aus der Sicht Georg Forsters«, in *Georg Forster in interdisziplinärer Perspektive*, hg. von C.-V. Klenke, Berlin 1994, 147–156.

——, »Zurück zu den Affen oder über die natürliche Güte des Menschen. Rousseaus Kulturkritik und die Folgen«, in *Georg-Forster-Studien* II, Berlin 1998, 79–107.

Kühn, M., *Kant. Eine Biographie*, München 2003.

Lepenies, W., »Historisierung der Natur und Entmoralisierung der Wissenschaften seit dem achtzehnten Jahrhundert«, in *Natur und Geschichte*, hg. von H. Markl, München/Wien 1983, 263–288.

——, »Georg Forster als Anthropologe und als Schriftsteller«, in ders., *Autoren und Wissenschaftler im 18. Jahrhundert. Linné—Buffon—Winckelmann—Georg Forster—Erasmus Darwin*, München/Wien 1988, 121–153.

Lichtblau B., »Vorstellungen über Kannibalismus. Ein wissenschaftstheoretischer und -geschichtlicher Diskurs über Anthropophagie bei Michel de Montaigne, Johann Gottfried Herder, Georg Forster, Adolf Bastian, Leo Frobenius und Ewald Volhard. Eine quellen- und ideologiekritische Studie«, Dissertation im Fach Ethnologie an der Universität Wien 1997.

Liesegang, T., »Das Skandalon der Revolution. Zu Georg Forsters ›Parisische Umrisse‹ im Kontext zeitgenössischer Öffentlichkeitstheorien«, in *Georg-Forster-Studien* XI/2, Kassel 2006, 497–519.

Mahlke, R., »Georg Forster und die Bibliothek seines Vaters—ein Desiderat der Forsterforschung«, in *Georg-Forster-Studien* IV, Kassel 2000, 152–156.

Mathy, H., »Die letzten Aktivitäten Georg Forsters als Mainzer Universitätsbiblio-thekar«, in *Gutenberg-Jahrbuch* 1979, 319–324.

May, Y., »Pluralität und Ethos in Georg Forsters Anthropologie«, in *Georg-Forster-Studien* XI/2, Kassel 2006, 335–357.

——, »›Ganz ist Herder doch mein Mann nicht‹. Georg Forster und Johann Gott-fried Herder. Brennpunkte einer Freundschaft unter Vorbehalt«, in *Georg-Forster-Studien* XII, Kassel 2007, 231–256.

——, *Georg Forsters literarische Weltreise. Dialektik der Kulturbegegnung in der Aufklä-rung*, Berlin/Boston 2011.

Mazauric, C., »Art. ›Terreur‹ «, in *Dictionnaire historique de la Révolution française*, hg. von A. Soboul, Paris 1989, 1020–1025.

McLynn, F., *Captain Cook. Master of the Seas*, New Haven/London 2011.

Meyer, A., »Mechanische und organische Metaphorik politischer Philosophie«, in *Archiv für Begriffsgeschichte* 13 (1969) 128–199.

Muschg, A., *Von einem, der auszog, leben zu lernen. Goethes Reisen in die Schweiz*, Frank-furt am Main 2004.

Neugebauer-Wölk, M., »Das Rote und das Schwarze Buch—zur politischen Symbolik der Mainzer Jakobiner«, in *Die Publizistik der Mainzer Jakobiner und ihrer Geg-ner. Revolutionäre und gegenrevolutionäre Proklamationen und Flugschriften aus der Zeit der Mainzer Republik (1792/93). Zum 200. Jahrestag des Rheinisch-Deutschen*

Nationalkonvents und der Mainzer Republik. Katalog zur Ausstellung der Stadt Mainz im Rathaus-Foyer vom 14. März bis 18. April 1993, Mainz 1993, 52–68.

Neumann, M., »Philosophische Nachrichten aus der Südsee. Georg Forsters *Reise um die Welt*«, in *Der ganze Mensch. Anthropologie und Literatur im 18. Jahrhundert*, hg. von H.-J. Schings, Stuttgart/Weimar 1994, 517–544.

Peitsch, H., *Georg Forster. A History of His Critical Reception*, New York 2001.

———, »Forster und Goethes ›Prometheus‹«, in *Georg-Forster-Studien* XII, Kassel 2007, 353–368.

———, »Georg Forsters ›deutsche‹ Kommentierung englischer Reisebeschreibungen über den Pazifik«, in ders. (Hg.), *Reisen um 1800*, München 2012, 251–264.

Pickerodt, G., »Georg Forster als politischer Schriftsteller«, in *Georg-Forster-Studien* IV, Kassel 2000, 1–16.

———, »Der ›gesellschaftliche‹ Schriftsteller‹. Friedrich Schlegels Blick auf Forsters intellektuelle Physiognomie«, in *Georg-Forster-Studien* VI, Kassel 2001, 51–65.

Rasmussen, D., »Georg Forster als gesellschaftlicher Schriftsteller und seine Beziehungen zu Schlegel und Schiller«, in ders. (Hg.), *Der Weltumsegler und seine Freunde. Georg Forster als gesellschaftlicher Schriftsteller der Goethezeit*, Tübingen 1988, 189–200.

Reemtsma, J. Ph., »Mord am Strand. Georg Forster auf Tanna und anderswo«, in ders., *Mord am Strand. Allianzen von Zivilisation und Barbarei*, Hamburg 1998, 21–83.

Reichardt, R., »Die visualisierte Revolution. Die Geburt des Revolutionärs Georg Forster aus der politischen Bildlichkeit«, in *Georg-Forster-Studien* V, Kassel 2000, 163–227.

Reinalter, H., »Johann Georg Forster als Freimaurer und Rosenkreuzer«, in *Georg-Forster-Studien* I, Berlin 1997, 67–83.

———, »Johann Georg Forsters Revolutionsverständnis«, in *Georg-Forster-Studien* VIII, Kassel 2003, 207–218.

Revel, J., Art. »Die Große Angst«, in *Kritisches Wörterbuch der Französischen Revolution*, hg. von F. Furet und M. Ozouf, 2 Bde., Frankfurt am Main 1996, Bd. I, 110–121.

Richter, D., *Goethe in Neapel*, Berlin 2012.

Ritter, Ch., »Darstellungen der Gewalt in Georg Forsters Reise um die Welt«, in *Georg-Forster-Studien* VIII, Kassel 2003, 19–51.

Saine, Th. P., Georg Forster, New York 1972.

Sanches, M. R., »Dunkelheit und Aufklärung—Rasse und Kultur. Erfahrung und Macht in Forsters Auseinandersetzungen mit Kant und Meiners«, in *Georg-Forster-Studien* VIII, Kassel 2003, 53–82.

Sangmeister, D., »Das Feenland der Phantasie. Die Südsee in der deutschen Literatur zwischen 1780 und 1820«, in *Georg-Forster-Studien* II, Berlin 1998, 135–176.

Sauerland, K., »Zwei Revolutionsauffassungen—Goethe einerseits, Forster andererseits«, in *Georg-Forster-Studien* XIII, Kassel 2008, 215–233.

Schabert, T., *Natur und Revolution. Untersuchungen zum politischen Denken im Frankreich des achtzehnten Jahrhunderts*, München 1969.

Scheuer, H., »›Apostel der Völkerfreiheit‹ oder ›Vaterlandsverräter‹«?—Georg
  Forster und die Nachwelt, in *Georg-Forster-Studien* I, Berlin 1997, 1–18.
Schmied-Kowarzik, W., »Der Streit um die Einheit des Menschengeschlechts.
  Gedanken zu Forster, Herder und Kant«, in *Georg Forster in interdisziplinärer
  Perspektive*, hg. von C.-V. Klenke, Berlin 1994, 115–132.
Schöne, A., *Schillers Schädel*, München 2002.
——, »Unter aller Würde der Societät«, in *Jahrbuch der Akademie der Wissenschaften
  zu Göttingen 2009*, Berlin/New York 2010, 441–442.
Segeberg, H., »Georg Forsters Ansichten vom Niederrhein. Zur Geschichte der Reise-
  literatur als Wissensspeicher«, in *Georg-Forster-Studien* V, Kassel 2000, 1–15.
Seibt, G., *Mit einer Art von Wut. Goethe in der Revolution*, München 2014.
Soboul, A., *Die Große Französische Revolution. Ein Abriß ihrer Geschichte (1789–1799)*,
  hg. und übersetzt von J. Heilmann und D. Krause-Vilmar, Frankfurt am Main
  ⁵1988.
Stammen, Th., »Reisen in die Hauptstadt der Revolution. Georg Forsters ›Parisische Umrisse‹ und Walter Benjamins ›Moskauer Tagebücher‹ als Medium politischer Ordnungsreflexionen«, in *Normative und institutionelle Ordnungsprobleme
  des modernen Staates*, hg. von M. Mols/H.-O. Mühleisen/Th. Stammen/B. Vogel,
  Paderborn/München/Wien/Zürich 1990, 291–310.
Steiner, G., »Jacobiner und Societät der Wissenschaften«, in *Filológiai közlöny: a Mag-
  yar Tudományos Akadémia, Modern Filológiai Bizottsága és a Modern Filológiai Társa-
  ság világirodalmi folyóriata*, 4 (1958) Heft 4, 684–693.
——, *Georg Forster*, Stuttgart 1977.
——, *Freimaurer und Rosenkreuzer—Georg Forsters Weg durch Geheimbünde. Neue For-
  schungsergebnisse auf Grund bisher unbekannter Archivalien*, Weinheim 1985.
——, »›Uns hat zu Männern geschmiedet die allmächtige Zeit‹«. Die Biographie der
  Beziehungen zwischen Goethe und Georg Forster, in *Goethe und Forster. Studien
  zum gegenständlichen Dichten*, hg. von D. Rasmussen, Bonn 1985, 7–19.
Stilett, H., *Von der Lust, auf dieser Erde zu Leben. Wanderungen durch Montaignes Welten.
  Ein Kommentarband anderer Art*, Berlin 2008.
Strack, Th., *Exotische Erfahrung und Intersubjektivität. Reiseberichte im 17. und 18. Jahr-
  hundert. Genregeschichtliche Untersuchung zu Adam Olearius—Hans Egede—Georg
  Forster*, Paderborn 1994.
Thamer, H.-U., *Die Französische Revolution*, München 2004.
Uhlig, L., »Georg Forster und Herder«, in Euphorion. Zeitschrift für Literaturge-
  schichte 84 (1990), 339–366.
——, »Georg Forsters Horizont: Hindernis und Herausforderung für seine Rezep-
  tion«, in *Georg Forster in interdisziplinärer Perspektive*, hg. von C.-V. Klenke, Berlin
  1994, 3–14.
——, »Georg Forster und seine deutschen Zeitgenossen«, in *Georg-Forster-Studien* I,
  Berlin 1997, 155–170.
——, »Zwischen Politik, Bellestristik und Literaturwissenschaft. Georg Forsters Bild
  in der Kulturtradition des 19. Jahrhunderts«, in *Georg-Forster-Studien* VI,
  Kassel 2001, 1–24.

————, »Theoretical or Conjectural History. Georg Forsters *Voyage Round the World* im zeitgenössischen Kontext«, in *Germanisch-Romanische Monatsschrift*. Neue Folge, Bd. 53 (2003), 399–414.

————, *Georg Forster. Lebensabenteuer eines gelehrten Weltbürgers (1754–1794)*, Göttingen 2004.

————, »Hominis historia naturalis—Georg Forsters Vorlesung von 1786/87 im Zusammenhang seiner Anthropologie«, in *Studien zur Wissensschafts- und zur Religionsgeschichte*, hg. von Akademie der Wissenschaften zu Göttingen, Berlin/New York 2011, 159–221.

————, »Der polyglotte Forster. Fremdsprachige Bekenntnisse im Zusammenhang seines Lebens«, in *Georg-Forster-Studien* XVIII, Kassel 2013, 135–178.

Veit, W., »Topik einer besseren Welt: Der Pazifik als Gegenwelt und Herausforderung Europas«, in *Georg-Forster-Studien* XI/1, Kassel 2006, 177–228.

Warneken, B. J., *Schubart. Der unbürgerliche Bürger*, Frankfurt am Main 2009.

Willms, J., *Tugend und Terror. Geschichte der Französischen Revolution*, München 2014.

Wuthenow, R.-R., *Vernunft und Republik, Studien zu Georg Forsters Schriften*, Berlin/Zürich 1970.

————, »Enzyklopädische Reisebeschreibung: Georg Forsters ›Ansichten vom Niederrhein . . . ‹«, in *Georg-Forster-Studien* I, Berlin 1997, 85–100.

Index